Stress Monitoring in the Workplace

Stress Monitoring in the Workplace

Kaare Rodahl

LEWIS PUBLISHERS
Boca Raton   Ann Arbor   London   Tokyo

Library of Congress Cataloging-in-Publication Data

Rodahl, Kåre, 1917–
  Stress monitoring in the workplace / Kaare Rodahl.
    p.   cm.
  Includes bibliographical references and index.
  ISBN 0-87371-655-8
  1. Job stress—Measurement—Methodology.   2. Work—Physiological aspects.
3. Industrial safety.   I. Title.
RC963.48.R63 1993
616.9'803—dc20   93-11815
                                                                                                  CIP

This book contains information obtained from authentic and highly regarded sources. Reprinted material is quoted with permission, and sources are indicated. A wide variety of references are listed. Reasonable efforts have been made to publish reliable data and information, but the author and the publisher cannot assume responsibility for the validity of all materials or for the consequences of their use.

Neither this book nor any part may be reproduced or transmitted in any form or by any means, electronic or mechanical, including photocopying, microfilming, and recording, or by any information storage and retrieval system, without prior permission in writing from the publisher.

CRC Press, Inc.'s consent does not extend to copying for general distribution, for promotion, for creating new works, or for resale. Specific permission must be obtained in writing from CRC Press for such copying.

Direct all inquiries to CRC Press, Inc., 2000 Corporate Blvd., N.W., Boca Raton, Florida 33431.

© 1994 by CRC Press, Inc.
Lewis Publishers is an imprint of CRC Press.

No claim to original U.S. Government works
International Standard Book Number 0-87371-655-8
Library of Congress Card Number 93-11815
Printed in the United States of America 1 2 3 4 5 6 7 8 9 0
Printed on acid-free paper

Preface

The purpose of this presentation is to show how some of the well-established physiological techniques, combined with new, sophisticated electronic devices, can be used to transfer the basic work physiology laboratory from the academic setting into the field in order to accurately monitor and study the actual combination of stresses to which the worker is exposed. The advantage of this is obvious, in view of the fact that physical work—the way it is normally conducted, under the prevailing conditions of real life—cannot be entirely simulated in the laboratory.

On the basis of a general review of the physical, mental, and environmental stresses of work, some of the available techniques and tools for the monitoring of these stresses will be described, and a series of examples of such monitoring will be given. It is by no means meant to give a complete or comprehensive picture of this field, for this would not be possible; this field is in very rapid development, and it would be impossible for one person alone to master it all. This review will be based primarily on my own experience, supplemented by recent trends in the development of problem areas and the possible application of new technology.

It should be kept in mind that this is a way of supplementing basic academic research by applying its results and by extending it into the field.

The Author

Kaare Rodahl, M.D., D.Sc., obtained his M.D. degree in 1948 and his D.Sc. degree in 1950, both from Oslo University in Norway. He obtained his U.S. medical degree in 1957.

Dr. Rodahl formerly served as a Special Consultant to the U.S. Air Force (1949). He also served as the Chief of the Department of Physiology (1950–1952) as well as the Director of Research (1954–1957) at the Arctic Aeromedical Laboratory in Fairbanks, Alaska. From 1952 to 1954, he worked at Oslo University as an Assistant Professor of Physiology. In 1957, he became the Director of Research at Lankenau Hospital in Philadelphia, and in 1960, he was made an honorary member of the Staff and Faculty of the U.S. Army Command and General Staff College, Fort Leavenworth, Kansas. He returned to Oslo in 1965 to serve as the Director of the Institute of Work Physiology and, starting in 1966, as a Professor at the Norwegian College of Physical Education; he retained both positions until 1987. He was awarded the Royal Norwegian Order of St. Olav in 1988.

Dr. Rodahl has participated in many Arctic and Antarctic scientific expeditions. He has authored some 200 scientific publications in the fields of nutrition, metabolism, stress, environmental physiology, work physiology, and the physiology of exercise. He has also written numerous popular articles covering the same subject matter. His scientific publications include *Textbook of Work Physiology,* 3rd edition (McGraw–Hill, New York, 1986), written with Professor P.-O. Åstrand, and *The Physiology of Work* (Taylor & Francis, London, 1989). He also edited a series containing *Bone as a Tissue* (McGraw–Hill, 1960), *Muscle as a Tissue* (McGraw–Hill, 1962), *Fat as a Tissue* (McGraw–Hill, 1964), and *Nerve as a Tissue* (Harper & Row, New York, 1966). He has also written numerous popular books on similar subjects.

Dr. Rodahl is presently retired and living in Oslo.

Contents

CHAPTER 1
Work Physiology Applied to the Worker in the Field 1
- Industrial Physiology in the U.S. 3
- Industrial Physiology in Norway 4
- Physical Fatigue 6
- Air Pollution 9
- Heat Stress in Norwegian Industry 9
- Industrial Cold Stress 16
- Problems Connected with Shift Work 18
- The Stress of Assembly Line Operators 22
- Work Stress of Norwegian Fishermen 22
- The Stress of Sailors at Sea 28
- Offshore Catering 31
- The Stress of Aircraft Pilots 31
- The Work Stress of Air Traffic Controllers 33
- The Work Stress of Airport Luggage Handling 37
- The Stress of Small Business Managers 38

CHAPTER 2
The Industrial Workplace as a Physiology Laboratory 39
- The Tools with Which to Do the Job 39
- Factors Which Affect our Capacity to
  Perform Physical Work 42

CHAPTER 3
Squirrel Logger–Compatible Sensors for the Ambulatory Counting of Heart Rate 49
- The Original Separate Heart Rate Recorder 51
- The Effect of Heat Stress on Heart Rate 55
- The Immediate Effect of Severe Heat Exposure on the Heart Rate 57
- The 8-Bit Squirrel with a Built-in Heart Rate Counter: The Eltek Special 60

CHAPTER 4
Squirrel-Based Ambulatory Logging of Muscle Tension as an Expression of Muscular Work Load 63
- The Use of the Myolog–Squirrel Combination for the Logging of Muscle Engagement 68
- Ambulatory Logging in the Assessment of the Claimed Relationship Between Muscle Tension and Neuromuscular Complaints 77
- Biofeedback 88

CHAPTER 5
The Monitoring of Heat Stress 91
- The Botsball Thermometer 92
- Continuous Recording of Environmental Temperature Indices 93
- Assessment of Heat Strain 94
    - Rectal Temperature 95
    - Skin Temperature 96
- Heat Stress and Heat Strain in a Typical Norwegian Aluminum Plant 97
- Heat Stress and Heat Strain in the Modern Glass Industry 99
- Effect of Adequate Fluid Intake 105
- Heat-Protective Clothing 106
- Protection of the Head 110
    - The Effect of Protective Helmets 114

CHAPTER 6
The Sensing of Humidity and Dust 121

CHAPTER 7
The Logging of Carbon Monoxide (CO) Exposure 125
- General Considerations 125
- Carbon Monoxide Concentrations in the Ambient Air 125
- Comparison of Three Available CO Sensors 127
- Assessment of the Actual CO Uptake by Estimating the
  COHb Content of the Blood with the Bedfont CO Monitor 130

CHAPTER 8
The Logging of Sulfur Dioxide (SO_2) Exposure 135

CHAPTER 9
Continuous Logging of Hydrogen Sulfide (H_2S) 141

CHAPTER 10
**A Simple Way of Assessing the Relative Concentrations of
Chemical or Organic Vapors in the Working Atmosphere** 145

CHAPTER 11
Further Possibilities of Sensor–Logger Combinations 153

REFERENCES 157

INDEX 163

Acknowledgments

I would like to take this opportunity to express my gratitude to my publisher, Jon Lewis, who encouraged me to write this book. I would also like to express my appreciation to R. Brandtzæg, L. D. Klüwer, and J. Cook for their support of some of the work reported in this book. I would also like to thank my colleagues and coworkers who have helped me in collecting the reported data or who gave me permission to use some of their observations. Above all, I would like to acknowledge T. Guthe, H. Nes, R. Karstensen, J. Meyer, L. Bjerke, L. H. Nitter, J. Grønli, A. Lie, L. Leuba, A. Kulsrud, R. Bjørklund, P. Søstrand, A. Rodahl, K. Maltun, T. Bottolfsen, and T. Eklund.

Finally, I would like to thank my wife, Joan Rodahl, for her help in preparing the manuscript.

*To Joan,
my faithful collaborator*

CHAPTER 1

Work Physiology Applied to the Worker in the Field

As students of physiology a generation or more ago, we were largely taught how the human body functioned at rest. The physiology of the body exposed to the stress of work is a subject of more recent decades.

I entered the field of applied work physiology by coincidence. In the late 1940s and early 1950s, physiologists on the whole tended to be more practical and field-oriented, partly as a consequence of World War II. At that time, there was a real need for a better understanding of the nature and limitations of human performance under stress and under adverse environmental conditions. This resulted in publications such as Adolph et al. (1947): *Physiology of Man in the Desert;* Newburgh (1949): *Physiology of Heat Regulation and the Science of Clothing;* Lehmann (1953): *Praktische Arbeitsphysiologie;* Floyd and Welford (1953): *Fatigue;* Burton and Edholm (1955): *Man in a Cold Environment;* Brouha (1960): *Physiology in Industry;* and Åstrand and Rodahl (1970): *Textbook of Work Physiology.* Much of this work was the result of physiologists observing and studying the problems where they occurred, seeking solutions in the field, if possible, or bringing the problems back into the laboratory for solution, and subsequent transformation of the results back to the field.

In my own case, it all started by spending a winter with Arctic trappers in northeast Greenland, 1939–40 (Figure 1-1), for the purpose of studying the toxic effect of Polar bear liver, which turned out to be caused by its huge content of vitamin A (Rodahl and Moore, 1943; Rodahl, 1949). This led me into the field of nutrition, and gave me some arctic experience as well. In turn, this was the basis for my being called to Alaska by the U.S. Air Force in 1949 to help them decide what survival rations to use for their Arctic air crews flying the B-52 bombers, carrying atomic bombs over the North Polar Basin. The successful solution to this problem (Rodahl, 1950) resulted in my being appointed as the head of the Physiology Department of the Arctic Aeromedical Laboratory at Ladd Air Force Base in Fairbanks, and subsequently as Director of Research at that laboratory. This was at a time—long before the era of intercontinental missiles—when Arctic

FIGURE 1-1. A laboratory for vitamin analysis was established in a trapper's cabin at Revet in northeast Greenland, in 1939.

warfare was a matter of high priority, with particular emphasis on the soldier's ability to perform and to survive in the cold. The scientific pursuit of the problems involved turned out to be typical examples of applied work physiology involving the task of moving the research laboratory into the field, since the real conditions involved, including a combination of cold and prolonged physical activity under varied conditions of snow and terrain, could not very well be simulated in the laboratory.

The first question we set out to answer was why the Eskimos, native to the Arctic, were able to get along in the Arctic environment better than we did. This involved a systematic study of the physiology of the Eskimos living in their natural habitat, eating their normal diet, and engaged in their normal activities, including hunting, moving camp, etc. This necessitated establishing rather sophisticated laboratory facilities in the field, in remote Eskimo villages, or in tents among nomadic Eskimo tribes (Figure 1-2).

Our main finding was that an Eskimo gets along better in the Arctic than does a New Yorker for the same reason a New Yorker gets along better in New York than does an Eskimo. It is a matter of the individual getting used to and adapting to a way of life compatible with the prevailing conditions of the area in question (Rodahl, 1954).

Clearly, studies of this type could not be done in any basic academic laboratory; the natural circumstances, including the living and working conditions encountered in the subject's natural habitat, could not by any means be simulated in a laboratory.

With the introduction of intercontinental missiles, the emphasis on man-based military operations in the Arctic gradually declined. Consequently, less money became available for cold-weather research, and especially cold-weather physiology. Hence there was a noticeable decline in scientific activity in this field, including diminished recruitment of young talents. Eventually, this led to the closing of the Arctic Aeromedical Laboratory at Ladd Air Force Base.

INDUSTRIAL PHYSIOLOGY IN THE U.S.

My own Alaskan field activities came to an end in 1957 when I became associated with the Lankenau Hospital in Philadelphia as director of its Division of Research. Here I had the privilege of working with a number of outstanding scientists, both American and European. I was also involved in a wide spectrum of research projects, both basic and applied, which formed the basis for my subsequent applied work physiology field studies. The projects included basic metabolic studies in cooperation with Steven M. Horvath and others (Rodahl et al., 1962), and a series of bed rest studies as part of the space research program on the physiological effect of gravity and inactivity (Rodahl et al., 1964).

One of our ambitions at the Lankenau Hospital Division of Research was to develop an extensive program of applied work physiology in American industry. This was based on the recording and logging of pertinent physiological data on workers involved in industrial operations. The activity was initiated with the help of Theodor Hettinger from the Max Planck Institute of Work Physiology in Dortmund, Germany and Per-Olof and Irma Åstrand from the Swedish Institute of Work Physiology in Stockholm, Sweden.

It soon became apparent, however, that this task was not an easy one. We started by approaching some of the managers of some of the industries which we thought would be of interest. We were, however, flatly told that they were not interested in any observations or recordings being made in their plants for fear of the negative consequences of any unfavorable conditions which might be revealed. We then approached the workers but were told that they did not wish to serve as subjects or to allow any measurements to be made on them for fear of any negative findings which might jeopardize their continued employment. They simply feared that if any health or functional impairment of handicap were detected they might lose their job. Finally, we succeeded in circumventing all of their objections by explaining that our research team might be able to improve their physical working capacity by testing and training them and to help them by showing how they could accomplish their job with a minimum of expended energy, thus having more energy left over at the end of the working day for the enjoyment of their leisure time with their families and friends. This they understood and accepted, and this led to a work physiology study of the operators at an assembly line operation at a gasoline pump company in Maryland during December 1958. The results indicated, as could be expected, that the greater the level of physical fitness of the operator, the less evidence of fatigue or deterioration of

FIGURE 1-2. (A) A set of laboratories were established in tents at the nomadic Eskimo camp site at Anaktuvuk Pass in the Brooks Range in Alaska.

physical work capacity occurred during the working day, emphasizing the value of physical training for the purpose of improving physical fitness. But the study also revealed that the operators spent approximately 40% of their working day resting and preparing for work. The data suggested that it would be possible to considerably increase the overall productivity of the assembly line operation without increasing the load of the men beyond reasonable physiological limits. This could be achieved by reorganizing the work schedule and the distribution of the work tasks, and dividing the rest time more equally among the operators on the line (Hettinger and Rodahl, 1960).

INDUSTRIAL PHYSIOLOGY IN NORWAY

At Lankenau, we had the opportunity to test some of the physiological instruments developed by an electronics company for use during space flights, including a radio-transmitting heart rate recorder, permitting wireless transmission of heart rate over considerable distances. At that time it was already well established that the heart rate is a most useful indicator of the physical work load to which a person is exposed (Åstrand and Rodahl, 1986). The heart rate, generally speaking, increases linearly with the work load (Figure 1-3).

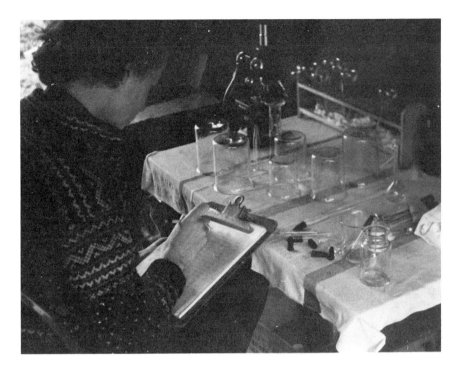

FIGURE 1-2. (B) In one of the tents, a metabolic laboratory was established for the collection and preparation of blood and urine samples, as well as for the weighing of the food intake.

This wireless heart rate transmitter was one of the instruments which I took back with me to Norway in 1965, when I became Head of a newly established Institute of Work Physiology, and subsequently also the Head of the Department of Physiology at the Norwegian College of Physical Education in Oslo. Here the wireless transmitted heart rate recorder was used to monitor the heart rate of members of the Norwegian speed skating team, in order to determine the percentage of their maximal heart rate taxed during different stages of the race, acceleration in the curves, and skating on the rinks' straight tracks. In a similar series of studies, a special pressure sensor, which was developed by the U.S. Navy to measure hydraulic pressure against the propeller of war ships under different speeds and settings, was used to record the pressure under the foot of speed skaters when skating in the curve and along the straight part of the rink, in relation to speed and acceleration.

The Norwegian Institute of Work Physiology was established as a public institution under the Department of Labor, for the purpose of contributing "to the maintenance of optimal health and working capacity of the industrial workers and to contribute to the adjustment of the working environment to suit the workers." In accordance with these objectives, a variety of projects were performed in different kinds of Norwegian industry (on land, at sea, and in the air). The purpose was to monitor work stress (physical, mental, and environmental) in working

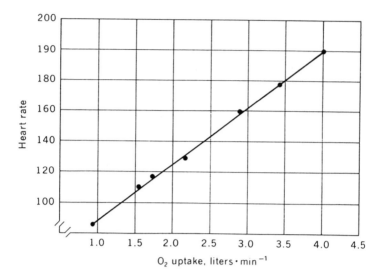

FIGURE 1-3. The heart rate, generally speaking, increases linearly with the work load. (From Åstrand, P. O. and K. Rodahl, 1986, *Textbook of Work Physiology*, 3rd ed., McGraw-Hill, New York. With permission.)

individuals, while they were working at their usual places of work. Most of these projects were developed at the request of the industry itself, and were represented by the health department, the management, or the workers' union.

Physical Fatigue

One of the earlier requests received was from a mechanical factory which, at the time, was engaged in the production of drive shafts for the Swedish Volvo automobile engines. These drive shafts, weighing 7.5 kg each, had to be lifted from the assembly line into the tempering oven to be hardened. Some of the workers wanted to know if they were exposed to a safe work load, considering the fact that they were lifting several tons of metal in the course of a working day. The task was further complicated by the shafts being covered with a thin layer of oil, making them so slick that none of the workers were able to hold the shaft by one hand in a vertical position. The only way they could manage was by grasping the shaft as it appeared on the assembly line with one hand, and swinging it in a perpendicular manner from the assembly belt into the oven.

A small temporary physiology laboratory was established in a separate room adjacent to the production hall, including instruments and equipment to measure the work load in relation to work capacity in each of the four workers involved in this operation, and to assess any objective physiological evidence of muscular and general physical fatigue. This included assessment of maximal working capacity by measuring maximal oxygen uptake by standard procedure. In addition, the heart rate was recorded at fixed submaximal work loads on the cycle ergometer at the beginning and at the end of each work day for 5 days running. The physical work load was determined by measuring the oxygen uptake in each of the subjects while working, using the Douglas bag method. This was repeated at the beginning and at the end of each work day for 5 days. In addition, the heart rate during work was recorded every other minute throughout the entire working day by the previously mentioned radioelectrocardiograph, transmitting the heart rate from the worker to the receiver some 20 m away. Blood samples for the analysis of the level of blood lactate were taken from the finger tip at the beginning and at the end of work, everyday. The rectal temperature was recorded at the beginning and at the end of each working day. The maximal muscle strength of both the right and left hand grip was measured with the aid of a hand dynamometer at the beginning and at the end of each working day. Finally, the body weights of the subjects were recorded on a balance with an accuracy of 20 g before and after the work shift in order to assess the subject's state of hydration. Food and fluid ingested and urine and stools eliminated were recorded on an accurate laboratory scale (for further description of the methods refer to Åstrand and Rodahl, 1986).

The results of these studies showed that the maximal oxygen uptake in liters per kilogram of bodyweight was within the acceptable limits for all of the subjects (41 to 52 mL O_2/kg b.w., Nilsson et al., 1970). The subjects taxed, on the average, 25% of their maximal work capacity when working. This relative work load was the same at the beginning and at the end of the working day (24.9 vs. 25.1%). This load is well within the limits of a reasonable physical work stress. This conclusion was supported by the results of the recording of the heart rate while working in each of the four subjects.

The recording of the heart rate during the submaximal cycle ergometer exercise (work pulse over 130 beats/min) before and after the work day, in five successive days, revealed no increase in the heart rate under these standardized conditions at the end of a day's work, nor as a consequence of a 5-day work week (Figure 1-4). The changes in the recorded rectal temperatures during the working day were no more than could be explained by the circadian rhythmic changes (0.3°C) and did not indicate any significant increase in body heat content due to elevated energy metabolism. As expected the blood lactic acid levels remained within the resting values, supporting the above-mentioned findings related to the levels of energy metabolism of the working subjects.

The recording of the muscle strength as assessed by the grip strength of the hand used to lift the drive shafts revealed higher values at the end of the working day than at the beginning of the day, and increasing values from day to day (Figure 1-5). Whether or not this could be the result of a training effect may be discussed, but the results certainly do not indicate any deterioration of muscle strength in the involved muscle groups as a consequence of overloading or fatigue.

On the whole, the body weight remained unchanged in all four subjects during the working day. This indicated that under the thermal conditions of the study (23 to 24°C) the subjects replaced their weight loss due to sweat loss (970 g on the average) by fluid intake of the same amount.

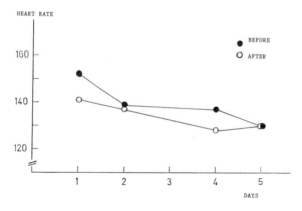

FIGURE 1-4. Mean heart rate of four workers at a machinery factory at a fixed work load on a cycle ergometer, recorded before and after the work day. The higher value before the first day may be due to the subjects having had a substantial meal less than an hour before the test.

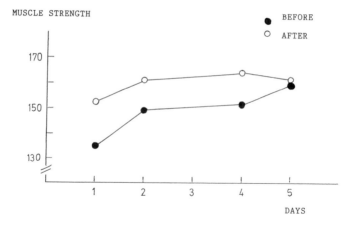

FIGURE 1-5. Muscle strength (units) as assessed by a hand grip dynamometer. Mean values of four subjects, before and after work for 5 days.

On the basis of these findings, we had to conclude, much to the workers' disappointment, that the physical work load to which the examined workers were exposed, at the time of the study, was within the limits considered reasonable for physical work. Furthermore, on the basis of the physiological observations made, we were unable to detect any objective recordable evidence of physical fatigue as the result of a day's or week's work under the existing conditions.

We have since come to the conclusion that physical fatigue during submaximal work is a most difficult parameter to objectively assess by the measurement of physiological parameters. This, for the time being, leaves us no choice but to continue to rely on subjective assessments with the aid of questionnaires and subjective assessment scales.

Air Pollution

There was a comprehensive occupational health survey of an aluminum production plant on the west coast of Norway, which particularly emphasized the effect of fluoride exposure. We were asked to assess the pulmonary function in some of the fluoride-exposed production hall operators, compared to some of the nonexposed office workers. The measurement of the forced expiratory volume ($FEV_{1.0}$) in 25 workers, showed that the pulmonary function, as indicated by the FEV, was significantly higher at the end of the work week than at the beginning of the week, which followed a weekend off work. Since tobacco smoking was not permitted in the production hall of the plant at that time, it was suggested that the increased pulmonary resistance after a weekend off work might be due to excessive tobacco smoking during the weekend when the examined workers were off work. There was no difference in the FEV between fluoride-exposed production hall workers and nonexposed office workers (Jahr et al., 1971).

Heat Stress in Norwegian Industry

At the request of the management and representatives of the Workers Union in a magnesium plant in the south of Norway, we studied the work load and the heat stress exposure of some of their workers operating electrically powered vehicles, tapping the finished metallic magnesium from the pots, and transporting it to the foundry. This study was done in order to help them decide the number of persons needed to operate each vehicle. The workers themselves maintained that two men were needed for reason of safety, one for driving and one for helping. The management felt that one was enough to do the job.

The heat stress to which the operators were exposed was measured in terms of the Wet Globe Temperature index (WGT) by the Botsball thermometer (Botsford, 1971). The WGT can easily be converted to the Wet Bulb–Globe Temperature index (WBGT) by a simple formula. (For references see Rodahl and Guthe, 1988.) The heat strain was assessed by the continuous recording of skin and rectal temperature from a small Oxford Medilog tape recorder carried on the subject. The same logger also recorded the heart rate continuously from three precordial electrodes, as an index of physical work load. Sweat loss was assessed by recording the body weight before and after work, and by the weighing of food and fluid intake, and the eliminated stools and urine.

On the basis of this study, we concluded that it made very little difference, in terms of safety and work load, whether the job was done by one or two persons. We did point out, however, that if two operators were used, the one who was not driving should stay behind the vehicle to avoid being exposed to the radiant heat from the pot. The nondriving operator stood on the exposed side of the cart in the past (Rodahl and Huser, 1976).

The results of this study caused the health and safety manager of the plant to express his concern about the heat stress exposure of some of his workers in

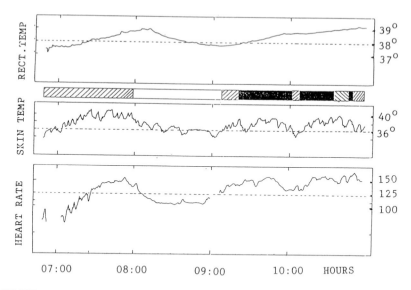

FIGURE 1-6. Slag removed in a magnesium plant, causing rectal temperatures well above 38°C.

general. He, therefore, asked us to carry out a survey of the thermal environment and its effect on the operators at several different work places of the plant. We were specifically asked to compare the heat stress at the old part of the plant, with that encountered at the part of the plant where a new type of closed-in electrolysis production pot was installed and used.

Our findings revealed considerable heat exposure, even at the new, closed-in pots (Figure 1-6). This was greatly enhanced by the fact that the operators elected to work uninterruptedly, as intensely and as long as necessary to get the job done as quickly as possible, instead of working for limited periods interrupted by brief resting periods in a cool environment. The workers were then able to spend the rest of the working shift relaxing in the canteen or reading newspapers in their restroom. The result was that their rectal temperature rose well above 38°C (Figure 1-7A). The same operators were persuaded to do the same work on a different schedule, working 20 min and taking 10-min breaks in the cooler environment outside. The same amount of work was done in about the same total amount of time, but the heat strain was much less, the rectal temperature staying well below 38°C all the time (Figure 1-7B).

In 1973, Kloetzel et al. published the results of a study in Brazil, and suggested a connection between industrial heat exposure and the development of hypertension. Based on these results, the head of the health department of one of our major industrial companies in Norway, asked us to make a survey of the

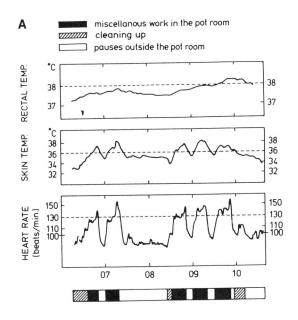

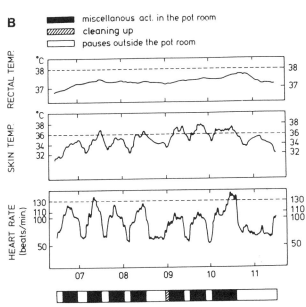

FIGURE 1-7. Worker in magnesium plant, slag removal. (A) Original work schedule; (B) modified work schedule (10 min work, 10 min cooling-off break).

heat stress and its effects at a ferroalloy plant in southern Norway. Two series of studies—one in the winter and one in the summer—were made in 1978 and again in 1979. The heat exposure in terms of Botsball temperature was higher than the recommended upper limits of about 25°C WGT in all the work areas

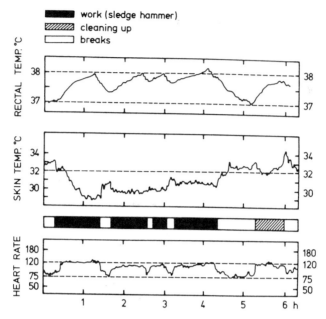

FIGURE 1-8. Heart rate and body temperature in a manual worker engaged in the breaking-up of finished metal with a sledge hammer in the cool environment of a ferroalloy plant.

examined in the furnace room. Rectal temperatures in excess of 38°C, which had been recommended as the ceiling value for internal body temperature in heat-exposed industrial workers, were observed in all but two subjects (Magnus et al., 1980). Similarly, a rectal temperature above 38°C was observed in one of the manual workers not exposed to ambient heat (Figure 1-8). This supported the physiological concept that an elevated central body temperature above 38°C is a common desirable experience in physical workers and in athletes, causing improved physical performance without any harmful effects (Åstrand and Rodahl, 1986).

Only about half of the fluid lost through sweating was replaced by fluid intake during the working shifts (Magnus et al., 1980). The degree of hypohydration expressed in percentage of the body weight was 1.15% on the average. This is close to the level at which objective signs of impaired endurance and other pathophysiological signs are observed (Åstrand and Rodahl, 1986).

The results of the blood studies revealed elevated plasma renin concentrations during heat exposure (Magnus et al., 1980). This was interpreted as an indication of reduced kidney blood flow. The renin elevation also occurred in the manual workers not exposed to heat but engaged in strenuous physical work. Thus, the cause of the reduced blood flow could be either heat or physical work, or both. In addition, the study indicated some degree of hemodilution during the working shift in the heat-exposed workers and emphasized the importance of heat acclimatization (Egeland et al., 1981).

Systolic and diastolic blood pressures were measured in a standardized manner in 50 heat-exposed and 50 non–heat-exposed workers in the plant and showed no significant difference. Nor was there a significant difference the measurement of blood pressure in ten subjects during and after the working shift (Magnus et al., 1980). A subsequent extensive 6-year follow-up study of the blood pressure and of the incidence of hypertension in the workers of this plant (Erikssen et al., 1990) supported the finding that exposure to heat stress is not associated with the development of high blood pressure. This was previously suggested by Kloetzel et al. (1973). For further details see Rodahl and Guthe (1988).

This project was extended by a similar study at one of our modern aluminum plants in the south of Norway. Here studies were carried out twice a year for 2 years (1979 to 1980) (Rodahl, 1981). Surprisingly, there was only about 10°C difference between winter and summer in the ambient temperature in the plant, but, as in the ferroalloy plant, the ambient temperature was considerably above the recommended upper limits at a number of job locations.

The workers in this plant were, on an average, exposed to the heat of the potroom for only 44% of the total working shift. The rest of the time was spent moving about, or in the canteen drinking coffee, talking, reading, or playing cards. The workers preferred, as is often the case in Norway, to work hard to get the work done as quickly as possible, so as to be able to rest that much longer. The undesirable result is often long uninterrupted periods of excessive heat stress, instead of more frequent but shorter periods of exposure, interspersed with brief cooling-off periods outside the potroom. As an example, the burner cleaner, illustrated in Figure 1-9, spent three 1-hour periods actually cleaning the burners in the potroom. The rest of the time he spent outside the potroom in the canteen, where the temperature was comfortably cool. The time actually spent by the pot, on the whole, varied from 30 to 130 min per shift.

In most of the heat-exposed workers, the rectal temperature exceeded 38°C. In one case, it was excessive for more than 60% of the observation period. When exposed to the potroom temperature, the subject's rectal temperature rose gradually, leveling off in about 45 min. The skin temperature reacted much faster, and may serve as a good indicator of how close the subject is to the heat source. It was observed that profuse sweating in most cases started at a skin temperature measured at the inside of the thigh of about 36°C.

The highest sweat rates which we observed were close to 1 L/hour. In several cases, they exceeded 1% of the body weight in the course of the working shift, indicating significant negative fluid balance. One worker consumed as much as 6 L of fluid in the course of his working shift.

The blood studies verified the findings of the ferroalloy industry. They showed a lower hemoglobin value at the end of the shift compared to the values before the shift, indicating hemodilution (Jansen et al., 1982). Those who had the greatest sweat loss also had the greatest drop in sodium concentration in the blood. In all cases, however, the values were within the normal range. In every case, by the time the subject appeared on the working shift the next day, the sodium loss was replaced. It was therefore concluded that there was no need for taking salt tablets.

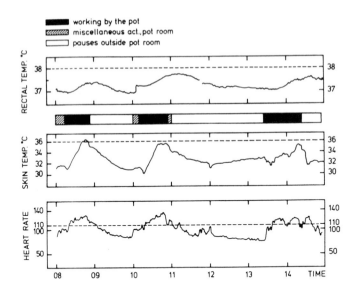

FIGURE 1-9. Work and rest pattern in a burner cleaner in an aluminum plant.

As a consequence of these findings, the health and safety committee of the plant established a working group to transform our findings into practical improvements, including the availability of proper drinking water in adequate quantities.

At the request of colleagues involved in the teaching of environmental health to engineering students at the University of Trondheim, in 1974 we conducted a survey of the nature and magnitude of heat stress to which the furnace operators were exposed at a Norwegian cement factory, in connection with the introduction of a system of remote control of the furnace operations.

Continuous heart rate recording in four operators, both at work and during their time off work, revealed, on the average, a moderate work load of some 19% of their maximal aerobic capacity, seldom exceeding 50%. The level of energy expenditure was even less during their leisure hours, as is evident from Figure 1-10. Evidently, much of the elevated heart rate during work may be due to mental or emotional stress caused by visitors, telephone calls, etc., since the level of physical activity, on the whole, was rather moderate with occasional brief spells of physical effort.

In a subsequent study a year later, in the same plant, the effect of fluid intake and the state of hydration on the heart rate were studied in six heat-exposed workers. The heart rate was recorded at a fixed work load of 100 watts on a cycle ergometer, before and after the working shift. This was repeated twice in each subject, always on the same shift, 1 day without the intake of fluid, and 1 day with the intake of 2 L of fluid in the course of the working shift. They all had a lower heart rate when fluid was taken than when they worked in the heat without taking fluid. The mean heart rate was 132 beats/min without fluid, and 121 when fluid was taken, the difference being statistically significant (Rodahl, 1975a).

HEART RATE

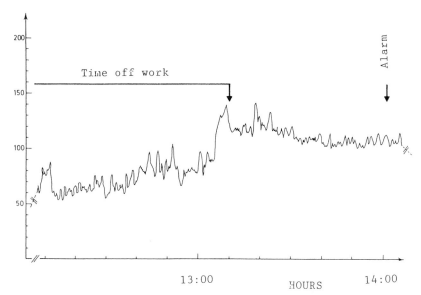

FIGURE 1-10. The heart rate in a cement furnace operator during work and leisure hours.

Similar findings were made in two heat-exposed spray painters in a kerosine stove production plant in Oslo. The heart rate at a fixed work load (600 kpm/min) was 8 to 14 beats/min higher when the subjects were working without fluid intake, than when they consumed 1750 mL fluid during the shift (Rodahl, 1975b).

In 1971, we were asked to assess the heat stress at a Norwegian paper mill, as part of an industrial hygiene survey, with particular reference to the working environment in the drying room. Two older individuals, who had cardiovascular problems, spent 2 hours a day cleaning the room for paper debris. The rest of the working shift were engaged in different outdoor jobs, such as attending to the garden, cleaning, etc.

Our survey revealed environmental temperatures in excess of 65°C. In one of the subjects, we recorded a drop in blood pressure to 90/60 at the end of 1 hour of work. In the other subject, who was 53 years old, we recorded a heart rate of 170 beats/min at the end of the first hour of work. This was probably close to his maximal heart rate. The sweat loss exceeded 1 L in 2 hours. It was concluded that the subjects in question were not suited for this particular job. In their cases, the job definitely represented a heart hazard. It was suggested that eventually the task be performed by mechanical devices. In our judgment, the job demand exceeded the normal human physiological limits of healthy individuals at best, let alone individuals with cardiovascular limitations, as was the case here (Rodahl et al., 1971).

In June 1984, a labor dispute developed in one of the major calcium carbide production plants in Norway. The problem involved whether or not it was safe to reduce the number of men from two to one for operating the tapping vehicle and tapping the finished product from the large, closed calcium carbide furnaces. It was finally decided by both parties involved to leave it to us to decide based on the measurement of the actual heat and work stress involved.

A total of 12 subjects were studied during the actual performance of the operations involved, both as operators of the tapping vehicle sitting inside a cooled cabin and as helpers, which also involved the strenuous job of tidying up after the tapping operation had been completed. Each subject was followed during the entire working shift. Measurements included continuous logging of the environmental heat stress by a Botsball thermometer, rectal and skin temperatures as an expression of heat strain, and heart rate as an expression of the combination of work load and heat stress with the aid of miniature Vitalog loggers.

Our measurements revealed quite clearly that considerable heat stress was involved. It was as great or greater than that previously recorded in the Norwegian melting industry, with Botsball temperatures up to 40°C. The total sweat loss amounted to 2 to 3 L in the course of the work shift. When working as the helper, the work and heat stress was much greater than when operating as the driver. The mean heart rate was up to 50% of the heart rate reserve, i.e., maximal heart rate less resting heart rate, and greatly elevated body temperature (see Figure 1-11 as an example). The work rate was exceptionally high: it used only 25% of the work shift for breaks and resting periods, which is considerably less than was the case in a Norwegian aluminum plant which was examined about the same time and where 40% of the work shift was used for breaks and rest periods (Rodahl, 1981).

On the basis of the data collected and the observations made, we concluded that it would be inadvisable, from a physiological point of view, to reduce the number of operators per tapping vehicle to one person (Rodahl et al., 1984). This was accepted by both parties concerned.

INDUSTRIAL COLD STRESS

One of the few industrial cold stress studies which we have been asked to carry out was a study of some Arctic coal miners in one of the Spitsbergen coal mines in 1979. Contrary to coal mines in the rest of the world, the temperature in the Spitsbergen coal mines is quite low, due to the permafrost. Because of the geological conditions—almost horizontal sedimentary layers and coal seams only 70 to 110 cm thick—the miners have to work lying on the ground. In order to get to the coal face, the workers have to crawl several hundred yards. The work is performed in a lying, half-sitting, or squatting position for two sessions of approximately 3 hours each, in each shift period. The temperature in the mine is –2 to –4°C all the year round, and the workers have always complained of finding it difficult to keep their feet warm.

In collaboration with the health department of the mining company, the actual work stress was assessed in four of the miners (Alm and Rodahl, 1979). They were studied for 24-hour periods, both during work in the mine and during time off and

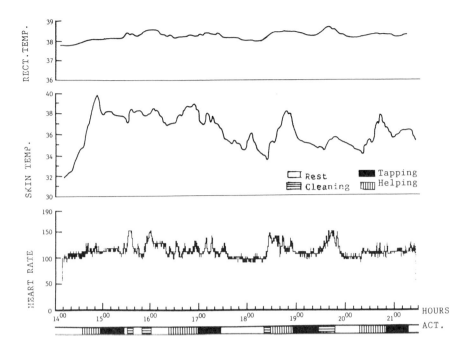

FIGURE 1-11. Heart rate and body temperature in a tapper in a calcium carbide production plant.

sleep. The study included (1) assessment of maximal work capacity (based on the recording of heart rate during submaximal cycle ergometer exercise); (2) assessment of physical work load (based on the continuous recording of heart rate with the aid of a shielded Oxford Medilog miniature portable magnetic tape recorder); (3) assessment of thermal stress (based on continuous recording of rectal and skin temperature by the same Medilog recorder); and (4) the assessment of general stress response (based on the analysis of urinary catecholamine elimination).

The estimated physical work load in the mine, which was quite similar for all four subjects, corresponded, on the average, to about 30 to 40% of their maximal work capacity (Figure 1-12). This is considerably higher than work loads commonly encountered in most industries, where they seldom exceed 25% of the maximal work capacity. It is evident from Figure 1-12 that this type of mining operation may impose some rather unique types of stress, as in the case when, at the onset of the work shift, the miner crawls along the narrow passage, dragging a box containing 50 kg dynamite tied to his leg, causing his heart rate to approach 165 beats/min. The work load of these coal miners is comparable to that of coastal fishermen. The levels of urinary catecholamine elimination of the fishermen equaled those observed in the coal miners (Figure 1-13).

The rectal temperature ranged from 37.5 to 38.5°C during work. In spite of the high rectal temperature, the skin temperature of the thigh dropped in two of the subjects to about 28°C (Figure 1-14). Thus, our observations supported the

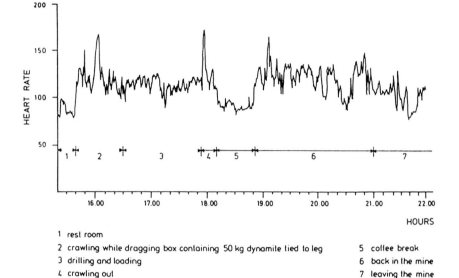

FIGURE 1-12. Heart rate in a coal miner in Spitsbergen during an afternoon shift. (From Alm, N. O. and K. Rodahl, unpublished, 1979.)

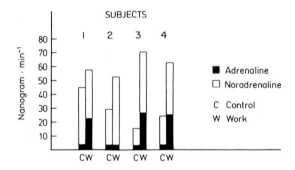

FIGURE 1-13. Mean urinary catecholamine elimination (adrenaline and noradrenaline) in four Spitsbergen coal miners during work, compared with night values. (From Alm, N. O. and K. Rodahl, unpublished, 1979.)

miners' complaints of cold feet, a problem which under the existing circumstances could only be remedied by using properly insulated trousers and boots.

PROBLEMS CONNECTED WITH SHIFT WORK

At the request of the management of a large iron-production plant in Norway in 1974, we made an extensive study of the physiological effects of the type of shift work practiced at this particular plant. The reason for their concern was partly

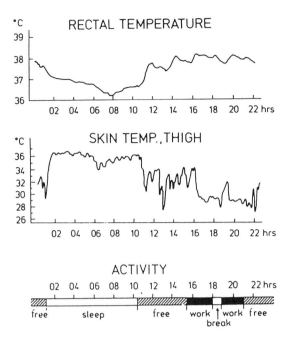

FIGURE 1-14. Body temperature (rectal temperature and skin temperature recorded at the medial side of the right thigh) in a coal miner in Spitsbergen during work and time-off. (Modified from Alm, N. O. and K. Rodahl, unpublished, 1979.)

the large number of workers needing rehabilitation or treatment for problems thought to be associated with the work schedules practiced, and which, in the final analysis, was a schedule chosen by the majority of the workers themselves. The problems associated with shift work was a fairly common one, even in a country like Norway, where 150 different companies at that time were based on around-the-clock shift work, i.e., morning, afternoon, and night shifts.

Based on the available literature, it appeared that the social implications of shift work, and the disturbing effect of shift work on family life, represented a greater problem than did the physiological effect. Surprisingly it appeared that absence due to sickness was much lower in shift workers than among day workers, but the reason for this was unclear. It was suggested that some of the untoward effects of shift work might be due to disturbance of the biological circadian rhythms of the working individual (for further references, see Rodahl, K., 1989).

In an attempt to elucidate some of these questions, we performed a series of work physiology studies of groups of workers at different work places in the iron works, both during winter and during summer. In one series of investigations, six shift workers were studied throughout an entire shift cycle, lasting about a month. In another series of studies, a group of shift workers agreed to work continuously on night shift uninterrupted for 3 weeks, including weekends and holidays. The purpose of the latter series was to find out whether or not continuous night shifts might develop a physiological adjustment of the circadian rhythm to night work, and if so, how long it would take for such a shift in a circadian rhythm to take place.

The recorded parameters included maximal work capacity (expressed by maximal oxygen uptake measured on a cycle ergometer) and determination of the physical work load (by direct measurement of the oxygen uptake during the actual work performed) supplemented by indirect assessment based on the continuous ambulatory recording of the heart rate during work compared with the heart rate during a series of known work loads on a cycle ergometer. The urinary elimination of adrenalin and noradrenalin was determined both during work and during rest. The rectal temperature was recorded before, during, and after work. This was done in order to reveal possible changes in the circadian rhythm, based on the fact that the rectal temperature, which is affected by the circadian rhythm, normally is lowest during the night, rising in the course of the day, and dropping again in the late afternoon. Similarly the heart rate was recorded at fixed submaximal work loads on a cycle ergometer, at a fixed time before, during and after work, in search of circadian rhythmic changes.

In the regular shift workers, it was observed that both the heart rate and the stress hormone–elimination levels were highest during the first night shift, after the subject returned to work from a 5-day free period (Vokac and Rodahl, 1975).

In the four workers who worked night shifts continuously for 3 weeks running, their deep body temperature declined during the first night shift as it does in a sleeping individual. There was a changing trend in the rectal temperature towards a smaller drop during the night shifts that followed, but even at the end of 3 weeks the rectal temperature still did not rise during work in the night. On the basis of these observations, it was concluded that it does not appear possible to attain a phase shift in the circadian rhythm during a 3-week continuous night shift schedule (Figure 1-15).

The stress hormone analyses revealed considerable individual differences in response to shift work. This may possibly be taken as an indication of individual differences in the physiological ability to cope with or tolerate shift work. The stress hormone study revealed quite clearly that the worst part of the practiced shift work schedule was to start on a night shift immediately after a 5-day accumulated free period. From our observations, it appeared that the prolonged time off was often used for a variety of intensely concentrated activities, such as building, traveling, hunting, fishing, etc., with a minimum of sleep. This caused some degree of sleep deprivation by the time the first night shift was due to start. This was particularly the case with the young workers (Vokac and Rodahl, 1975).

In a subsequent study, we had the opportunity to pursue the problem of circadian rhythm disturbances by studying the effect of crossing time zones in two officers on board a supertanker en route from South Africa to South America (Rodahl, 1980). During the voyage, the ship's clock was put back 1 hour every other night. In the first mate (who worked 4-hour shifts and had 8 hours free) and in the chief engineer (who only worked during the day), the rectal temperature was continuously recorded every other 24-hour period for 2 weeks. In the officer working regular shifts, there appeared to be a slight shift to the left in the circadian

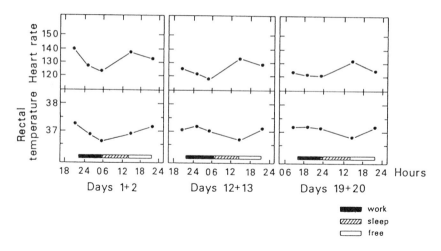

FIGURE 1-15. Mean heart rate at a fixed submaximal work load, and mean rectal temperature in four steel-mill workers during three continuous weeks of night shifts (Modified from Vokac, Z. and K. Rodahl, in "Experimental Studies of Shiftwork", Forskungsberichte des landes Nordrheim–Westfalen, Nr. 2513, 168-173, Westdeutscher Verlag, 1975. With permission.)

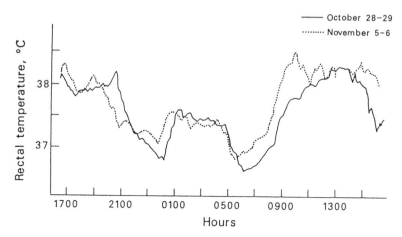

FIGURE 1-16. Rectal temperature (24-hour) of the first officer on board S/S Docecanyon en route from South Africa to Brazil, recorded at the beginning and at the end of the voyage. The ship's clock was put back 1 hour every other night. The record may indicate a slight shift in the circadian rhythm to the left (Rodahl, K., 1989).

rhythm, when comparing the temperature record for the first and the last day of the voyage (see Figure 1-16). This suggested a possible lag in the rhythm which disappeared during the subsequent days when the ship remained at the same latitude on the east coast of Brazil. In the chief engineer, who worked during the day only, and was free to sleep regularly during the entire night, no such phase shift was observed.

THE STRESS OF ASSEMBLY LINE OPERATORS

In 1974, we were asked to conduct an assessment of the stress involved in a company producing stoves in Oslo. The results of the study were compared with a similar study carried out at an assembly-line operation producing gasoline pumps in the U.S. in 1958 (Hettinger and Rodahl, 1960).

The Norwegian assembly-line operation proceeded more or less continuously throughout the work shift, with a 6-min break once every hour and nine operators working in a row. The total time used for rest and eating in the Norwegian plant amounted to 22% of the shift, compared to 27% in the American plant. In the Norwegian study, about 5% the time was used getting ready and preparing for the actual production line operation vs. 30% in the American study. The net result was that some 70% of the time was spent for the actual assembly task in the Norwegian plant vs. some 40% in the American plant. On the average, the physical work load taxed about 22% of the workers' maximal work capacity in the Norwegian case vs. some 25% in the American study (Rodahl, 1975b).

WORK STRESS OF NORWEGIAN FISHERMEN

At the initiative of the Norwegian Fishermens Association, and with the support of the Norwegian Ministry of Fisheries, we conducted a series of studies of Norwegian fishermen, including coastal as well as deep-sea fishing. This involved accompanying the fishermen at sea (sometimes in very small boats), logging data, and collecting samples under extremely difficult conditions, e.g., in high seas and rough weather. The data were analyzed in the laboratory on land.

The reason for this project was the need for objective physiological data as a basis for (1) the evaluation of a person's fitness for work in the different branches of the fishing industry; (2) early retirement; and (3) assessing the degree of disability in fishermen with disabling diseases.

Twenty-four full-time, professional fishermen engaged in common types of coastal fishing off Lofoten (i.e., handline, longline, net, and Danish seine) were studied for a total of 35 working days during the fishing seasons in 1971 and 1972. An initial study was carried out during the spring of 1971 under quite unfavorable weather conditions (Åstrand et al., 1973). That was followed by a supplementary study during the summer of 1971, under extremely favorable weather conditions. The final study took place in the spring of 1972, during which eight catch-handlers working ashore were also studied (Rodahl et al., 1974).

In a laboratory established ashore in a small fishing village (Svolvær), the maximal oxygen uptake was determined by the Douglas bag method using a mechanically braked cycle ergometer. This was also used to determine the heart

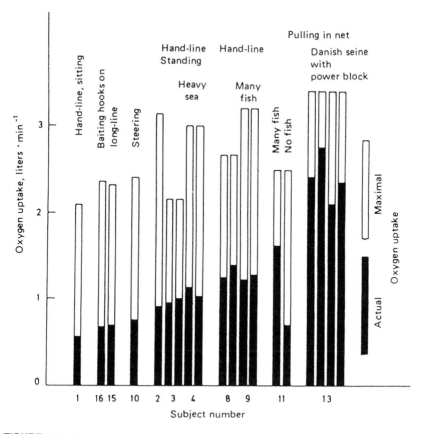

FIGURE 1-17. Measured oxygen uptake in typical fishing operations (Åstrand, I. et al., Scand. J. Clin. Lab. Invest. *31*:105, 1973. With permission.)

rate response to known submaximal work loads. Out at sea, the physical work load was assessed by the continuously recorded heart rate with a portable, battery-operated tape recorder, carried on the subject. In addition, the work load of some typical fishing operations was determined directly by measuring the subject's oxygen uptake using the Douglas bag method. The volume of the expired air was measured on board the fishing vessel by a dry spirometer. The collected samples of the gas were brought back and analyzed in the laboratory ashore. For the assessment of the total work stress to which the subject was exposed, urinary catecholamine elimination was determined in urine samples collected during the day and during the night. Urinary creatinine excretion was used as a check of the urine sample collection.

The energy cost of some of the typical fishing operations, in terms of actually measured oxygen uptake, is shown in Figure 1-17. It is observed that fishing with the handline in a sitting position (subject No. 1) represents the lowest energy expenditure, and was therefore suggested as a most suitable occupation for older individuals, or for those who had a reduced work capacity due to a medical or physical handicap. Pulling in the Danish seine by power block represented the highest work load, with an oxygen uptake of 2.0 to 2.7 L/min, corresponding to 60 to 80% of the subject's maximal work capacity (Subject No. 13). This activity, which occurred every hour or two, lasted for a few minutes at a time, and altogether lasted for some 20 min a day. In this kind of fishing operation, each man has to keep up with the rest of the team. This means that every member of the team has to be physically fit.

Of the different activities on board the different types of fishing vessels in the course of a working day, the mean energy expenditure amounted to about 1 L O_2/min, which is equivalent to 34 to 39% of the fishermen's maximal work capacity, and had occasional peaks up to 80%. On board small fishing vessels out at sea, some of this energy expenditure is due to the extra energy expenditure of counterbalancing the motion of the deck, requiring considerable muscular effort. In very rough seas, this may increase the level of energy expenditure by as much as 30% (Rodahl, K., 1989).

The study of the eight catch-handlers working ashore during a total of 12 working days showed a mean energy expenditure taxing some 34% of their maximal work capacity, which is surprisingly high. The peak values, however, were considerably lower and of much shorter duration than in the fishermen.

The average urinary excretion of stress hormones during the working day was remarkably high, especially in the fishermen. Here a nearly tenfold increase in the epinephrine excretion and a fourfold increase in norepinephrine excretion were observed during the day, as compared with the excretion during the night (Åstrand et al., 1973).

A separate examination of a few fishermen who received disability compensation revealed that although they were classified as disabled, their maximal physical work capacity, as assessed by tests on the cycle ergometer, did not differ markedly from the rest of the fishermen who were classified as being healthy.

The work stress associated with trawler fishing was investigated on board two medium-sized Norwegian stern trawlers, one operating off the coast of Labrador catching Greenland halibut, the other operating in the Barents Sea catching cod, haddock, etc. (Rodahl and Vokac, 1977a). The circulatory strain and the physiological work load were assessed by continuously recorded heart rate, in six subjects working in regular 6-hour shifts. In both trawlers, roughly one third of the time was spent working, one third resting, and one third sleeping. The actual work load to which they were exposed when working, on the average, taxed 30 to 38% of their heart rate reserve, with peaks up to 80%, corresponding to an oxygen uptake of about 1 L/min, which was close to the values observed in the coastal fishermen. The mean work load of the trawler skipper, on the other hand, whose work was entirely sedentary, corresponded to only 20% of his heart rate reserve. The average energy expenditure for a 24-hour period and the urinary catecholamine excretion rates in the fishermen on board the trawlers were definitely lower than in the coastal fishermen.

A similar study was performed aboard two longline fishing vessels. One of them was a modern Faeroe island trawler with a crew of 15, fishing Greenland

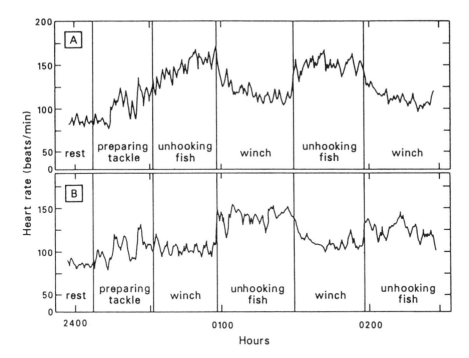

FIGURE 1-18. Heart rates in two deck hands (A and B) aboard a Norwegian longline fishing boat, fishing dogfish off the Orkney islands, changing places at regular intervals between unhooking the fish and operating the winch.

halibut off the coast of Labrador. The other was a Norwegian vessel with a crew of 11 fishing dogfish off the Orkney islands (Rodahl and Vokac, 1977b). Altogether five crew members were examined.

The findings were quite similar aboard the two vessels, and the work load in general of the same magnitude as that observed in deckhands aboard the trawlers. By far, the most strenuous operation was the unhooking of the fish as the longline was pulled in by the winch over the side of the ship, an operation which could not be endured for more than some 25 min at a stretch. A routine was therefore established, whereby the deckhand unhooking the fish and the one operating the winch, changed places at regular intervals (Figure 1-18). The urinary catecholamine excretion rates in these bank fishermen were of the same order of magnitude as those observed in the trawler fishermen. Periods of extremely high work stress were observed aboard the Norwegian vessel when the fishing was exceptionally good and the crew had to work around the clock without sleep for several days running. This was because the vessel at the time operated with a reduced number of crew.

Finally, three crew members (the skipper, a deckhand, and the cook) of a Norwegian purse seine fishing vessel catching capelin off the coast of Finnmark were studied. The study included a 24-hour continuous observation period during a fishing trip lasting 72 hours. Due to the remarkable efficiency of the methods developed for capelin fishing, the vessel was filled to capacity in a matter of a few hectic hours. Nonetheless, the circulatory strain observed in the skipper was fairly moderate, with a mean heart rate of about 90 beats/min. During the critical periods of the actual fishing operations, however, his heart rate reached peak values up to 155 beats/min. In the case of the cook, his average work load only taxed some 15% of his heart rate reserve. The 23-year-old deckhand, who served as the helper to the purse seine foreman, was subjected to exceptionally high work loads during the actual fishing operations, with peak loads exceeding 50% of his heart rate reserve lasting for a total of 100 min out of the 24-hour observation period (Rodahl, K., 1989).

From these studies it was evident that fishing may entail hard physical work. Yet, if properly organized with job rotation and an adequate number of crew members to share the work, it may, in some respects, become quite similar to some ordinary occupations ashore. The introduction of mechanization and automation, has greatly improved the efficiency of the fishing operations. However, this development has also led to an increased tempo, imposing a greater stress on the crew at times. This is especially the case in the older fishermen, who, on account of their declining physical work capacity, may find it difficult to keep pace with their younger colleagues, when the fishing operations are carried out by a team, as in the case of the longline, net or Danish seine fishing. Yet, the official Norwegian statistics show that almost half of our fishermen are over 50 years of age, and that the percentage of older fishermen is increasing.

During the studies of the longline and trawler fishing off the coast of Labrador, we had the opportunity to visit an Eskimo village on the Northwest coast of Greenland. A study, similar to that carried out on the fishermen, was performed on some Eskimo seal hunters (Vokac and Rodahl, 1976). The study was carried out in September 1973 in Kraulshavn, which at that time had a population of about 100 Eskimos. Eight Eskimo hunters volunteered to serve as subjects for the study. A laboratory was established in the Eskimo village. A Monark cycle ergometer, which we had brought with us, was used for the assessment of maximal work capacity and for the conversion of heart rate to corresponding work loads. Since bicycling is an exercise entirely unfamiliar to the native Eskimo, a great deal of time was used to teach the Eskimo subjects how to use the ergometer, and they were given ample time to practice. The subjects then exercised both at submaximal and maximal work loads. In the latter case, the subjects were encouraged to do their very best to keep their pace with a couple of Eskimo drummers beating their drums in time with the metronome. Heart rates were recorded on a conventional electrocardiograph. Pulmonary ventilation and oxygen uptake were measured by the Douglas bag method, using a dry spirometer and a micro–Scholander analyzer. Peak blood lactate concentration was determined from capillary blood collected from the prewarmed finger tip 5 to 6 min after the cessation of the maximal exercise.

In the field, the assessment of the circulatory strain was done in the usual manner, by recording the Eskimo subject's heart rate continuously on a portable battery-operated Hellige tape recorder which was carried on the person for 24-hour periods,

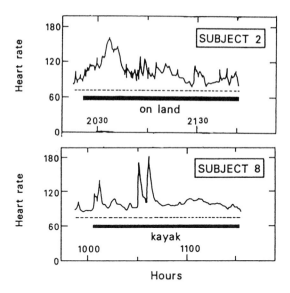

FIGURE 1-19. Continuous heart rate recordings in two Eskimo subjects during typical periods of activity, including seal hunting. The resting heart rates are indicated by dashed lines. (From Vokac, Z. and K. Rodahl, Nordic Council Arct. Med. Res. Rep., No. 16:16, 1976. With permission.)

during the Eskimo's usual activity, including seal hunting. In addition, urinary catecholamine output was assayed in urine samples collected from 0800 to 2000 hours, and from 2000 to 0800 hours.

As expected on the basis of our previous experiences with Eskimo subjects in Alaska, the maximal oxygen uptake, measured on the cycle ergometer, was exceptionally high in six of the Eskimo subjects (50 to 62 ml O_2/kg bw/min), and average in two of them (41-45 ml O_2/kg/min), according to generally accepted standards (Åstrand and Rodahl, 1986). These results were much higher than those reported earlier for Eskimos in the same community (Lammert, 1972). As a matter of fact, the mean maximal oxygen uptake in four of our subjects, who also served as subjects in the previous study, was about 50% higher than in the earlier study previously referenced.

The heart rate in one of the Eskimo subjects during a seal hunt in a kayak is illustrated in Figure 1-19. At times the Eskimo subjects remained sitting in their kayaks for hours at a stretch, waiting for the seal to appear, with only occasional slow paddling alternating with short bouts of extreme effort (subject No. 8). This, in our experience, appears to be rather typical for the Eskimo energy expenditure even when hunting.

The values of urinary catecholamine elimination, both during the day and during the night, were considerably lower than those observed in Norwegian coastal fishermen, indicating a comparatively low degree of stress in Eskimo hunters compared with those engaged in the modern fishing industry (for further details, see Vokac and Rodahl, 1976).

THE STRESS OF SAILORS AT SEA

At the request of the Norwegian Shipowner's Association, we became engaged in a research project concerning the physiological aspects of safety at sea, during the period 1976 to 1980 (Rodahl, 1980). Ninety-eight crew members of ten different ships were examined during regular voyages in different parts of the world, from the North Sea to the Persian Gulf, from Europe to the Far East. In each case, a field laboratory was established in one of the cabins on board. As a rule, at least two subjects were studied simultaneously. The duration of the stay aboard the ship varied from 1 to more than 4 weeks. The study included a general survey of the individual work places, supplemented by personal interviews with those working there. Detailed time–activity logs were kept for each subject, covering his activities, minute by minute.

A visualization of the subject's reaction to his work stress was made by the continuous recording of his heart rate by battery-operated, miniature recorders carried on the subject during the period of observation (Hellige, Oxford Medilog, and Vitalog). The relationship between heart rate and work load was established for each subject individually, with the aid of a cycle ergometer aboard the ship. Thus, the recorded heart rate could be translated into actual work load. In some cases, the subject's maximal physical work capacity was determined with the aid of the ergometer. In some cases, the blood pressure of the captain, during critical work situations, was recorded by a Bosomat II automatically inflated blood pressure cuff. This was supplied with a 4-m-long tube, enabling the captain to move about on the bridge and to perform his job without restrictions.

As an indication of the general stress reaction, the urinary elimination of epinephrine and norepinephrine was determined in urine samples collected aboard and later analyzed in the laboratory at home.

In crews exposed to heat, skin and rectal temperatures were recorded continuously over 24-hour periods with the aid of sensors connected to the Medilog recorder. The first and most extensive study in this series was a study on board a container vessel en route from Singapore to Panama via Japan. It transpired that the main results of this initial study were confirmed and supplemented by additional parameters in the subsequent studies.

The results showed that in spite of the limited facilities on board for physical training activity, the level of physical fitness of the sailors was not greatly different from that of comparable age groups in the population as a whole. However, it was somewhat lower in the officers than in the ordinary crew members. The physical work load, in general, was fairly similar to that of ordinary Norwegian industrial workers ashore (Figure 1-20). The heaviest physical work load, relatively speaking, was observed in women working in the mess and cleaning the cabins (Figure 1-21), in contrast to some of the cooks, who enjoyed an exceptionally light work load (Figure 1-22). Generally speaking, we found that the lower the rank, the greater the physical work load.

An interesting observation was that aboard small vessels, such as a tugboat operating at the oil fields in the North Sea, the almost constant motion, due to

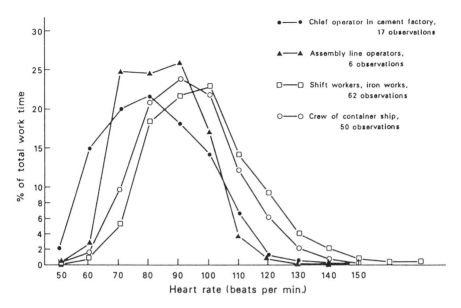

FIGURE 1-20. Work loads in different occupations, expressed in terms of heart rate recorded during an ordinary work day.

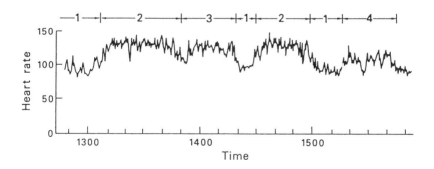

FIGURE 1-21. Physical work load in cabin attendant aboard M/S Bergensfjord: (1) break, (2) washing floor, (3) polishing floor, (4) odd jobs.

waves and wind, caused more or less constant elevation of the heart rate even when resting and during sleep. This appears to be because of the necessity of counteracting the motion of the ship. The mean heart rate during sleep on board the larger container vessel M/S Toyama was 65 vs. 74 on board the small tugboat T/B Tender Pull.

The work of the ship's officers, when sailing in open water, could be quite monotonous, especially during the night, when one could often observe a drop in the heart rate as a consequence of the circadian rhythm (Figure 1-23).

In captains of ocean-going vessels, maneuvering in and out of harbors usually caused an elevation of the captain's heart rate, indicating mental stress (Figure 1-24).

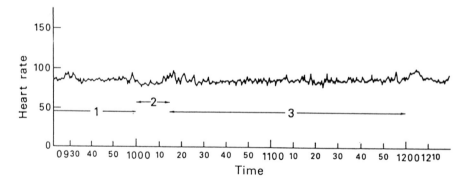

FIGURE 1-22. Physical work load of the cook aboard M/S Toyama: (1) odd jobs in the kitchen, (2) coffee break, (3) preparing lunch.

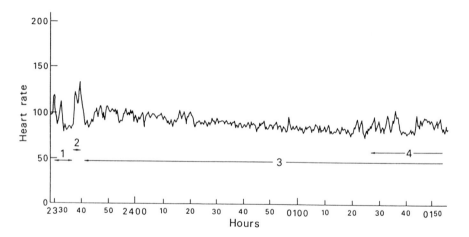

FIGURE 1-23. Heart rate in the second officer on night watch on the bridge of the M/S Toyama. The falling heart rate is due to the circadian rhythm: (1) in the cabin, (2) walking up to the bridge, (3) standing watch on the bridge, (4) worried about problems at home.

The same applied to the blood pressure (Figure 1-25). In some cases, this was associated with an increase in urinary stress hormone (catecholamine) elimination (Blix et al., 1979).

In the crew of ships operating in the Persian Gulf, assessment of heat stress revealed rectal temperatures exceeding 38°C in up to 40% of the entire work period (Figure 1-26). In the officers working outside on the deck, the work load exceeded 50% of their maximal work capacity for more than 50% of the time vs. only about 9% for the deckhands, on the average. These studies served to emphasize the importance of the use of modern air-conditioning systems and adequate fluid intake as an important measure of preventing the pathophysiological consequences of heat stress in sailors in the tropics.

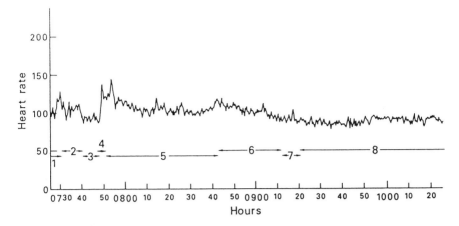

FIGURE 1-24. Evidence of mental tension in the captain of the M/S Toyama during the departure from Kobe harbor in Japan: (4) walking up to the bridge, (5) standing on the bridge with the coastal pilot, (6) in charge on the bridge after the coastal pilot had left the ship, (8) standing on the bridge in open water.

OFFSHORE CATERING

A private company, providing catering services for two of the offshore oil producing platforms in the North Sea, the Ekofisk and the Statoil, had a problem. Although the type of work, the number of people catered for, and the working conditions appeared to be quite similar on the two platforms, the absence from work due to sickness in the two groups of employees was drastically different, being much higher at the Statoil installation. The company's health department asked us to conduct a study of the level of work stress in relation to absence from work due to sickness. Unfortunately, due to the time it took to get started and because of a number of interacting circumstances, this particular study was never pursued. Instead, an elaborate project, involving a survey of the working environment, the work stress, musculoskeletal complaints, and sick absenteeism among the catering personnel at the Statfjord offshore oil field was undertaken (Westgaard et al., 1987).

One aspect of this study covered 42 catering employees studied for 24-hour periods, working a total of 8 hours a day every day for two consecutive weeks. From the results, it is evident that the work load, as assessed by the heart rate during the whole work day, was considerably greater than in a number of occupations ashore. An example of an exceptionally hard-working subject is presented in Figure 1-27.

THE STRESS OF AIRCRAFT PILOTS

At the request of the pilots of an airline operating small aircraft, such as the Twin-Otters in the northern part of Norway, we made a survey of the cockpit

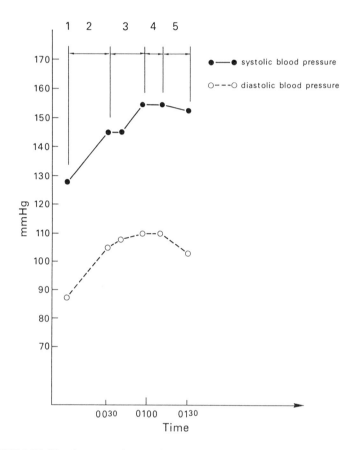

FIGURE 1-25. Blood pressure in captain of supertanker during departure from Angra dos Reis: (1) control values, (2) on the bridge, (3) waiting for the arrival of the tug boat, (4) leaving the port, (5) full speed ahead.

environment and the work stress of the pilots during some of their routine operations. The most striking feature of the cockpit environment, apart from the noise, was the uneven distribution of the temperature and how it was affected by changing altitude during take-off and landing (Figure 1-28). Another interesting finding was the stress of getting to work (especially the worry of the road conditions in the winter) in pilots who had elected to live in the remote countryside of the northern parts of Norway. The snowfall during the night might make it exceedingly difficult for the pilot to drive his car along the unploughed roads and to report to the airport before 4 o'clock in the morning (Figure 1-29), as opposed to a pilot who was given the opportunity of spending the night in a hotel within walking distance of the airport (Figure 1-30).

The most dramatic stresses we observed were those associated with landing on small airstrips located alongside steep mountains in strong turbulent airmotions (Figure 1-31). The purpose of these studies was to record reality

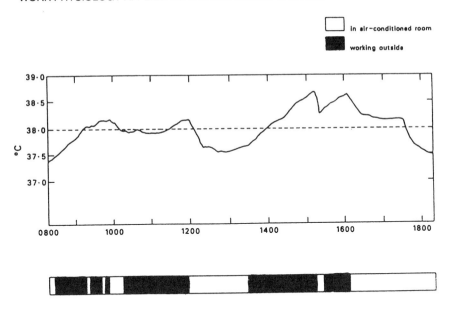

FIGURE 1-26. Rectal temperature in the chief officer of a cargo boat, M/S Tarn, in the Persian Gulf in the course of a working day (From Maehlum et al., 35-Rapportserien, Oslo, 1978. With permission.)

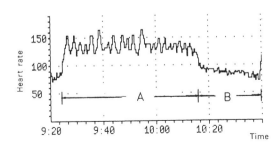

FIGURE 1-27. (A) A catering worker carrying bags containing clean bed clothes from the container up to the 1st floor, 12 bags at a time, a total of 15 trips; (B) break (Westgaard et al., Universitetsforlaget, Oslo, 1987. With permission.)

in order to gain an insight into the problems involved. Since the time of these studies, at least one fatal accident has occurred due to the turbulent conditions described above.

THE WORK STRESS OF AIR TRAFFIC CONTROLLERS

In 1981, the Norwegian air traffic controllers were engaged in a conflict concerning wages and working conditions. The controllers claimed that their level of responsibility was equal to that of the airline captains and should, therefore, be

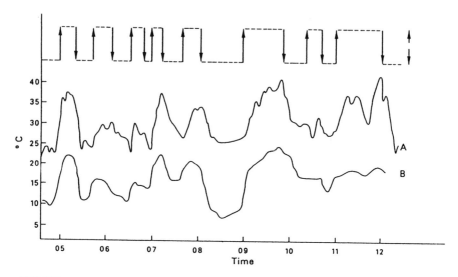

FIGURE 1-28. Air temperature in the cockpit of a Twin-Otter passenger plane at floor (B) and head (A) level. Arrows indicate take off (↑) and landing (↓).

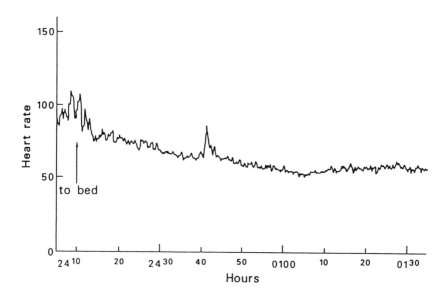

FIGURE 1-29. Heart rate in a commercial pilot, worried about getting to work on time due to a snowfall in the night.

WORK PHYSIOLOGY APPLIED TO THE WORKER IN THE FIELD

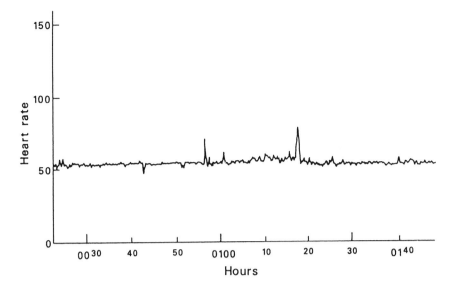

FIGURE 1-30. Heart rate in a commercial pilot (co-pilot to the subject in Figure 1-29), who spent the night prior to take-off in a hotel near the airport.

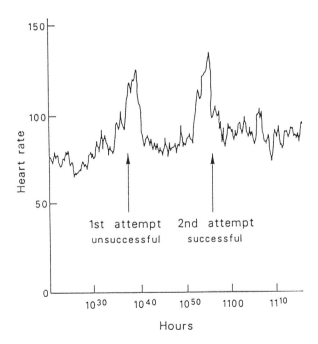

FIGURE 1-31. Heart rate response in commercial pilot landing a Twin-Otter aircraft at a small airfield under difficult weather conditions. He made one unsuccessful attempt but succeeded to land on the second attempt. He is not performing any actual physical work during the landing.

paid equally well. To emphasize their point of view, the air traffic controllers introduced a go-slow procedure drastically reducing the number of aircraft take-offs and landings, causing numerous delays and serious problems for the travelers, who developed a visible dislike for the controllers.

At the joint request of the Civic Aviation Agency and the Air Traffic Controller Union, a study was initiated in 1981. It was then repeated in 1982 when the conflict was solved, and then again in 1988. The purpose of the study was to survey the state of health with particular reference to stress in the air traffic controllers at the Oslo Airport (Rodahl et al., 1981). Altogether, 33 out of 46 air traffic controllers were examined. The study included a comprehensive medical examination with particular reference to blood pressure, measured by a standardized procedure, gastrointestinal disease, and risk factors such as blood lipids, etc. In addition, the heart rate, as an indication of autonomous nerve system response, was recorded for periods up to 24 hours (with the aid of miniature magnetic tape recorders), with the simultaneous recording of a detailed activity log. This enabled us to record how each subject reacted to different situations at work and during leisure hours.

The findings of this study supported the impression gained from previous studies: the work of the air traffic controllers may be demanding. Yet, our results did not indicate that they were subject to unreasonable levels of stress. The state of health appeared, on the whole, to be good; although symptoms and signs of gastrointestinal complaints appeared to be somewhat high (Rodahl, K., 1989).

Blood pressure and resting heart rate were examined by standardized procedures in all of the 33 air traffic controllers during the conflict in 1981, followed by a reexamination of 27 of them in 1982 when the conflict was solved, and then again in 23 of them in 1988 when there was no conflict (Mundal et al., 1990).

The mean systolic blood pressure of the 27 air traffic controllers in 1982 was 116 mmHg vs. 136 in 1981, a highly significant difference ($p < 0.001$). In 1988, the mean blood pressure was 121 mmHg in the 23 air traffic controllers examined on all three occasions. This is significantly lower than in 1981, only 5 mmHg higher than in 1982, and roughly equivalent to the expected increase due to increase in age. The mean diastolic pressure was significantly higher ($p < 0.001$) in 1981 (90 to 87) than in 1982 (75 to 73) and 1988 (81 to 78). The difference between 1982 and 1988 was what could be expected due to age. The resting heart rates were essentially unaffected. It should be noted that the Civic Aviation Agency representative responsible for the negotiations with the Air Traffic Controller Union also developed a transient elevation of his blood pressure during the conflict.

From these observations it appears that the elevated blood pressure observed during an occupational conflict (which lasted a few months) is reversible. The question remains how long such a transitory elevation of the blood pressure has to last in order to become permanent. On the basis of the data now available, it appears that the answer is a matter of years (Erikssen et al., 1990).

These observations suggest that the possible effect of occupational conflicts on the blood pressure of those engaged in the conflict should be kept in mind when

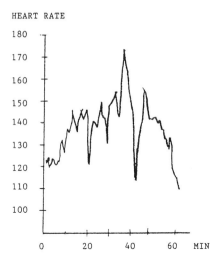

FIGURE 1-32. Luggage handler stacking newspaper inside aircraft in a squatting position.

considering blood pressure levels in occupational groups. In fact, it may explain some of the conflicting results reported in the literature concerning the blood pressure of air traffic controllers (Rodahl, K., 1989).

The message conveyed to the air traffic controllers themselves was that if they had to become engaged in occupational conflicts, the sooner they could get the conflict settled, the better.

THE WORK STRESS OF AIRPORT LUGGAGE HANDLING

As part of a comprehensive survey of the occupational health problems related to the operation of a major airline; in 1976, we were asked to conduct a series of work stress studies on personnel involved in the loading and unloading of luggage into and out of passenger aircraft at Oslo Airport. The study involved 10 of the regular luggage handlers, working in two shifts. The parameters recorded included the usual assessment of maximal work capacity, continuous recording of heart rate, and urinary catecholamine elimination (Rodahl, 1976).

In general, the work load appeared to be well within reasonable limits, ranging from 15 to 30% of the individual's maximal work capacity. The most striking feature was the work stress involved in the stacking of heavy bundles of newspapers, weighing 15 to 20 kg each, into aircraft in a squatting position. This sometimes demanded more than 50% of the individual's maximal capacity for as long as 30 min. uninterrupted (Figure 1-32).

Another source of stress was the struggle of getting to work in the morning. This was especially noticeable for those who had elected to live in a town far away

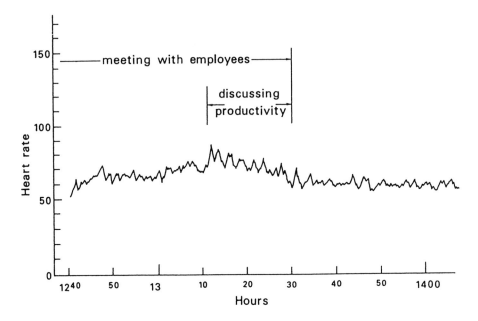

FIGURE 1-33. Heart rate reaction in a general manager during a discussion with his employees in a general assembly on the need for greater activity.

from the airport on the other side of a mountain ridge. Early in the morning, they chose to drive across the mountain on icy roads and through deep snow in the winter in order to arrive at work in time, rather than spending the night at the available facilities at the airport which were provided for such circumstances.

THE STRESS OF SMALL BUSINESS MANAGERS

Most studies concerning occupational health and aspects of applied work physiology have been done on workers. Far less interest has been devoted to the leaders of industry, especially the leaders of small and medium-sized industry, who, at times, may be working under a considerable amount of strain. For this reason, an attempt to assess the work stress in 21 directors of small and medium-sized companies (less than 150 employees) in different parts of Norway was made during the period 1982 to 1985 (Rodahl, K., 1989). Subjects were selected according to a table of random numbers. This study included around-the-clock logging of heart rate over 1 to 2 days in order to visualize some of the most pertinent sources of tension facing them, such as the emotional reactions of the manager during a discussion of productivity (Figure 1-33).

The general impression, from the study of this limited but fairly representative group of managing directors of small industrial companies in Norway, was that as a group, they were rather healthy, fit, and quite industrious (Rodahl et al., 1985).

CHAPTER 2

The Industrial Work Place as a Physiology Laboratory

For a couple of decades we, at the Norwegian Institute of Work Physiology, had devoted our resources to collecting data for Norwegian industry on their work and environmental stresses and on the effect of these stresses on their workers. Through these studies, the investigators learned a great deal about the problems involved in modern Norwegian industry. From all indications, industry itself gained very little in terms of practical insight into the nature of some of their own problems, about which they did not do anything. Part of this perhaps may be explained by the fact that it took so long for us to analyze the data collected and to inform the industry about our findings. This, in turn, was due to the complicated recording and computer equipment available at that time.

In 1985, it was decided that in Norway, as in some other industrialized countries, some of the control of the working environment should be transferred from public inspectors to the industry itself. This caused me to realize that instead of doing the measurements for the industry, we should have spent our time teaching them how to do these measurements on their own. This would have given them better insight into their own problems. Furthermore, it would have allowed them to use this information in their daily decision making in order to improve the working conditions, health, and well-being for their employees, and for greater productivity. Above all, it would have provided the industry with objective facts. From these data, management and labor jointly could have discussed what to do about the problems revealed, including what improvements to make. Based on visible facts, this might even have contributed to making the management and labor feel that they were in the same situation and had common interests.

THE TOOLS WITH WHICH TO DO THE JOB

This approach created the need for simple, easy-to-operate ambulatory loggers capable of logging several parameters simultaneously. These loggers had to be

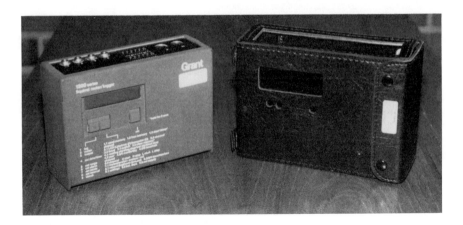

FIGURE 2-1. The Squirrel meter/logger with carrying case.

carried on the worker in order to record the parameters in question at the place where the person was working, and while the actual work was being performed. For this reason, the loggers had to be robust, but light, and they had to be able to store the recorded data in an internal memory for instantaneous display on a portable computer screen. This would allow everyone concerned, both labor and management, to observe the recorded results together and enable them to discuss the facts and what to do about them.

At that time, there were a variety of loggers available on the market. Most of them were limited to one or a few selected parameters. Recently, however, several more flexible, multichannel loggers have appeared, such as the Ramlog EI 9000 portable data logger (a.b.i. Data, Brussels, Belgium), the XT-107 Process Signal Logger (Colman Co.), the Vitalog Pocket-Polygraph (Vitalog Monitoring, Inc., U.S.), and the AMS-1000 Ambulatory Monitoring System (Consumer Sensory Products, Inc., Palo Alto, CA, U.S.A.).

The logger which was first brought to my attention had been available for some time and had proved to be quite reliable—the Squirrel meter/logger produced by Grant Instruments, Ltd., Cambridge, England (see Figures 2-1 and 2-2). It is capable of logging any parameter convertible to voltage or electric current. It has a large number of inputs and a high degree of flexibility. It is compact, light, and extremely simple to operate. It is set up more or less like a digital watch, using only three push buttons in connection with the liquid crystal display. The function of each button is actually printed on the logger, making it easy to use for nontechnical personnel. It can record several parameters simultaneously, related to both the working environment and the physiological responses to some of the environmental parameters, such as body temperature. It does this by using different sensors already available and which are compatible with the logger. It is a combined meter and data logger, i.e., it can be used to measure and immediately display data, such as ambient and body temperature, environmental gasses, electromyogram (EMG), heart rate, etc. At the same time, it can store the recorded

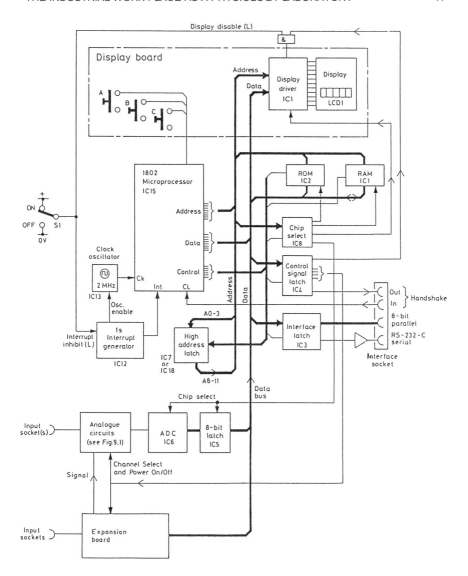

FIGURE 2-2. Schematic diagram of the Squirrel meter/logger.

data in its internal memory for immediate or subsequent transfer to a small, portable battery-operated PC, a battery-operated portable printer in the field, or a conventional office PC for data analysis, printout, or graphic display.

The 1200-series Squirrel is powered by six 1.5-volt alkaline batteries. It is specifically designed to use as little power as possible in the logging mode. With care, the batteries can be changed without loss of stored readings. The logger displays an indication of the remaining battery life. Because it is both robust and easy to operate, it is well suited for field use by nonacademic technical personnel.

Some of the models available have more than a dozen inputs, allowing a number of parameters to be recorded or logged simultaneously. This allows the investigator to record environmental parameters, such as ambient temperature, etc., as well as the subject's reaction to some of these parameters, such as core and skin temperature, heart rate, etc. The unit can display the data as they are being recorded, as a meter, while at the same time storing the readings in the electronic memory, as a logger. It is also able to display readings stored in it's memory. It is controlled by a microprocessor which is programmed to perform all the required measurements, display, and storage operations. Software programs are available for direct data transfer to any IBM-compatible personal computer for subsequent data analysis, display, and printout.

Both the conventional 12-bit Squirrel, as well as a special 8-bit version, are equipped with separate channels for the counting of the heart rate, based on the principle of counting the number of R-waves per minute. It is supplied with filters for the exclusion of artifacts or interference. In controlled laboratory tests, it has been found to be quite accurate and reliable. The 12-bit Squirrel meter/logger weighs about 1 kg and can be carried in a leather carrying case by a shoulder strap or by a belt around the waist (see Figure 2-1). The use of the Squirrel meter/logger for the purpose of recording and logging pertinent parameters with the aid of available sensors will be discussed later.

Since most of my recent experience is based on the Squirrel meter/logger, I hope that it is understandable that I have to limit the presentation of my data to the Squirrel meter/logger, and to describe it as a typical example, which to a large extent is transferable to any other similar logger or logger systems.

FACTORS WHICH AFFECT OUR CAPACITY TO PERFORM PHYSICAL WORK

With such a flexible, multisensor tool available, the next question which presented itself was what parameters do we want to sense. Since our primary objective was to record and assess some of the more obvious factors affecting the working individuals' ability to transform chemically bound energy in the food which is ingested into muscular contractions and physical work, we needed sensors capable of recording aspects of muscular function. This included not only muscle engagement but also the level of muscle tension, or force, based on the logging of the EMG from the muscles involved. Such sensors, compatible with the selected basic logger, were already available—as in the case of the Aleph Myolog sensor.

The transformation of chemically bound energy from the ingested food into mechanical energy in the muscle cell, in turn depends on the capacity of the service functions which deliver fuel and oxygen to the working muscle cell (Figure 2-3). This involves intake, storage, and mobilization of nutrients, i.e., fuel. It also involves the uptake of oxygen through the lungs and the delivery of this oxygen by the cardiovascular circulatory system to the muscle cell. In the muscle

THE INDUSTRIAL WORK PLACE AS A PHYSIOLOGY LABORATORY

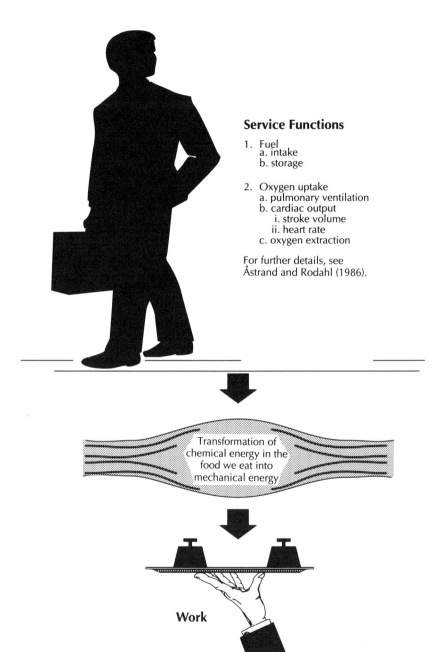

FIGURE 2-3. Factors affecting physical performance.

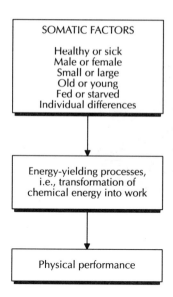

FIGURE 2-4. Somatic factors affecting physical performance.

cell, the fuel is oxidized for the purpose of providing energy for the muscle contraction (for further details, see Åstrand and Rodahl, 1986; Rodahl, K., 1989).

For the ambulatory assessment of the service functions supporting the energy-yielding processes in the muscle cell, sensors are available for the logging of pulmonary ventilation and oxygen uptake, as well as heart rate, as an indication of cardiac output.

The factors affecting the energy-yielding processes are not only physiological in nature. They include a variety of both psychological and clinical aspects, such as the worker's state of health and motivation. This is in addition to the nature of the work itself and the environment in which the work is being performed. In industrial work places where there are toxic agents in the ambient air, the uptake, and hence the noxious effect of the toxic agents, is dependent upon pulmonary ventilation and other physiological parameters which can be sensed and logged.

Of the somatic factors (Figure 2-4), one parameter which readily lends itself to an ambulatory sensor–logger system is the individual's state of physical fitness. This can be based on the heart rate at a fixed work load on a cycle ergometer or on a treadmill. (For references: Åstrand and Rodahl, 1986)

In the case of mental and psychosomatic factors (Figure 2-5), the individual's response to emotional or mental stress can be ambulatorily logged. This is based on the continuous recording of the heart rate as an indication of autonomous nervous system reaction to stressful situations. In our experience, based on an extensive series of studies, the recorded heart rate is a better indicator of mental or emotional stress than the urinary elimination of stress hormones (catecholamines). This is partly due to the difficulty of collecting complete urine samples in the field, apart from the fact that individual differences in urinary catecholamine

THE INDUSTRIAL WORK PLACE AS A PHYSIOLOGY LABORATORY 45

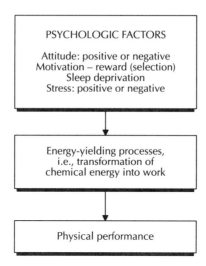

FIGURE 2-5. Mental and psychosomatic factors affecting physical performance.

levels may be greater than the differences due to different levels of stress exposure in the same individual (Rodahl, K., 1989).

When it comes to the work itself (Figure 2-6), many of its features may be characterized by different combinations of sensor–logger systems. The work load, or, in any case, the relative work load, may be assessed and directly recorded by using available sensors for the measurement of pulmonary ventilation and oxygen uptake, or indirectly recorded by using available sensors counting the heart rate. By using available electromyographic sensors, the nature and degree of muscular involvement can be determined, including the force developed, whether static or dynamic, continuous or intermittent, and the duration of the muscular involvement (Åstrand and Rodahl, 1986).

Of the environmental factors listed in Figure 2-7, suitable sensors are already available for the logging of all of them. In the case of temperature, there are suitable sensors for the logging of ambient temperature stress (air temperature, radiant temperature, Wet Bulb–Globe Temperature [WBGT], and Botsball or Wet Globe Temperature [WGT]), and its effect on the body temperature, both skin and central body temperature (rectal, ear, or esophageal). This may be of particular importance when considering the unique role played by the head in temperature sensation and control (for references, see Rodahl, K., 1989). In the case of the logging of the nature and degree of the air pollution in the working environment, the pulmonary uptake of polluted air is of particular importance. It is, therefore, of interest that, in some instances, the present sensor–logger system allows for simultaneous assessment of the polluting elements in the air and how much of this polluted air is taken up by the lungs and eventually transferred into the blood. In the case of carbon monoxide (CO), sensors are already available, allowing indirect assessment of the degree of hemoglobin-CO saturation. This is a matter of interest

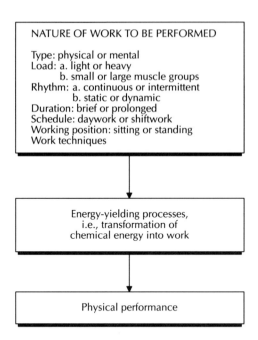

FIGURE 2-6. Factors, related to the work itself, affecting physical performance.

also for tobacco smokers, or as a means of encouragement for those smokers who are trying to quit smoking (Rodahl, K., 1989).

We shall now turn our attention to some of the available sensors which can be used in combination with loggers, such as the Squirrel, for the purpose of recording the previously discussed parameters that affect our health and performance. This will be illustrated and exemplified by some of the results obtained in pertinent field studies. In some cases, the practical implication of the findings will also be considered in the light of the previous discussion of the factors affecting our performance.

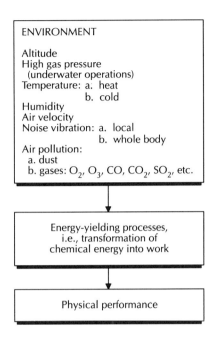

FIGURE 2-7. Environmental factors affecting physical performance.

CHAPTER 3

Squirrel Logger–Compatible Sensors for the Ambulatory Counting of Heart Rate

A person's heart rate is a most sensitive expression of his/her autonomous nervous system's reaction to stress of any kind, physical as well as mental. Under most conditions, excluding emotional tension, heat, etc., there is a linear relationship between work load and heart rate. The heavier the work load, the higher the heart rate (Figure 3-1). Similarly, emotional stress or tension is associated with an elevated heart rate (Figure 3-2). Under controlled conditions, therefore, heart rate can be used as an index of work stress.

There are a variety of ambulatory heart rate recorders available on the market, several of which we have used in the past. My own most recent experience is based on a miniature heart rate counter, the Eltek (produced by Eltek Ltd., in Cambridge, England). This Squirrel-based heart rate recorder is available in two different versions. The initial type consists of a separate unit for the heart rate counter and conversion from an electrocardiogram (ECG) to digital form. This type is plugged into a separate intake in the 1201 or 1202 series 12-bit Squirrel. The second version is the special 8-bit Squirrel with a built-in ECG pulse generator, known as the Eltek special. In both versions, they utilize the electrical discharges from the heart muscle, i.e., the electrocardiogram (EMG), as a method of counting the number of contractions of the heart per minute, i.e., heart rate. Actually, they count the number of R-waves per minute. Each version is supplied with filters for the exclusion of artifacts or interference. In controlled laboratory tests, we have found both types to be quite accurate and reliable.

For heart rate recording, the three disposable precordial electrodes are attached to the skin and connected to the power-pack–converter unit, i.e., ECG pulse generator connected to the Squirrel, and carried by a belt around the waist as shown in Figure 3-3.

It is recommended that the skin be cleansed, that the hair at the location of the electrodes be removed by a disposable razor, and that the skin be thoroughly

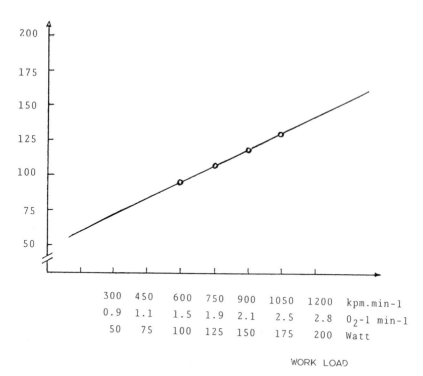

FIGURE 3-1. The heavier the work load, the higher the heart rate. (From Åstrand, P.-O. and K. Rodahl, "Textbook of Work Physiology", 3rd ed., McGraw-Hill, New York, 1986. With permission.)

rubbed with alcohol or a skin cleansing solution before attaching the electrodes. It is also strongly recommended that some extra electrode cream be put on the electrodes before they are attached to the skin. One of the sensing electrodes is placed at the top of the sternum, slightly to the left of the middle. The second is placed on the lower part of the chest, below the heart. The ground electrode is connected to the block input and the sensing electrodes are connected to the yellow and red inputs on the multi-way socket. It does not matter which of the two electrodes is attached to the red or to the black multi-way socket. The Squirrel is switched on and checked by activating function 2 (meter).

In our experience, it is advisable to secure the electrode connections by placing an adhesive sponge patch over each of the three electrodes. Also, secure the electrodes and their wire connections with the aid of a long elastic bandage to avoid loose contacts, especially in heat-exposed subjects who are apt to sweat a lot. The Squirrel, in it's leather case, is now ready to be attached to the subject by a strap or belt. Care should be taken to ensure that the press buttons on the Squirrel are not accidentally pressed when the subject is working. It may also be worth the effort to tape the wire connection to the Squirrel to avoid the plug being pulled out of the socket accidentally during intense activity.

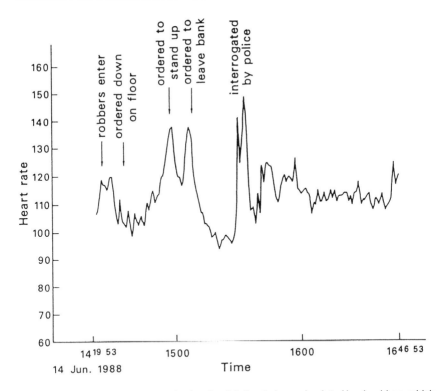

FIGURE 3-2. The heart rate reaction in a bank teller during a simulated bank robbery which included the taking of hostages (From Rodahl, K., 1989).

THE ORIGINAL SEPARATE HEART RATE RECORDER

In combination with a 12-bit Squirrel logger, the original heart rate recorder was used for the purpose of recording the physical work load of a ferroalloy plant worker in the South of Norway (February 1988) (see Figure 3-4). His job consisted of tapping the finished ferroalloy out of the pot three times in the course of the work shift. The rest of the time was spent in the canteen drinking coffee, smoking, and talking. The heart rate tracing, combined with a detailed time–activity log, revealed that a total of 3 to 4 hours of the entire work shift was spent working at the pot at an intense work rate exceeding 50% of his maximal work capacity, for a considerable part of the time—50% of the maximal physical work capacity is the equivalent of a heart rate of about 120 to 130 beats/min. Evidently, this type of work procedure which combined intense concentrated work effort with a prolonged, nonproductive recuperation or merely sitting in the canteen, was preferred by the workers. The workers did not seem to prefer extended work at a more reasonable rate and interrupted by more frequent, shorter rest periods. This is one of many examples of the use of the recorded heart rate, paired with time–activity records, not only to assess the magnitude of the work load as a basis for improvement of work procedures, but also as a general visualization of how the work is accomplished as a basis for a running evaluation of working efficiency.

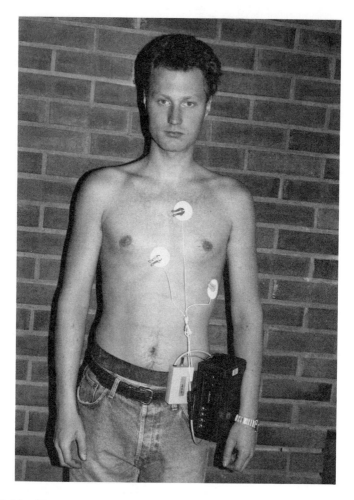

FIGURE 3-3. Photo of subject with precordial electrodes attached to heart rate recorder and Squirrel meter/logger.

Another application of this heart rate sensor combined with the 12-bit Squirrel logger was the recording of the heart rate as an expression of emotional strain in a bank teller during a simulated armed bank robbery (including the taking of hostages). The event was covered by a National Broadcasting team. It was repeatedly announced in advance and the involved bank employees were familiar with what was going to happen. They were told that at about 3 o'clock in the afternoon, just before closing time, two armed bandits would appear to rob the bank. They were all assured that no harm would be done.

One of the bank tellers volunteered to serve as a subject for our recording of his heart rate during the occasion. Electrodes were attached, the recorder was connected and fixed to his body by a belt around his waistline and covered by his

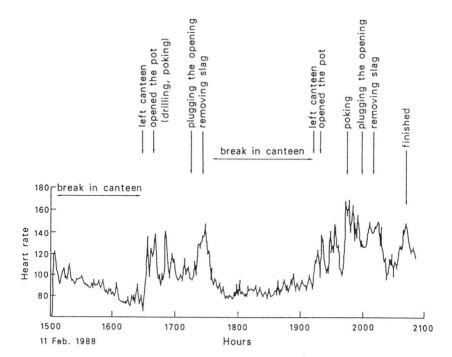

FIGURE 3-4. Heart rate in a worker in a ferromanganese production plant. Shows the effect of periods of strenuous physical work. (From Rodahl, K., 1989).

coat. The recorder was started well in advance of the event. The teller was left on his own. The investigator and the radio broadcasting team were discretely located behind a window in a separate room to observe the happening.

As is evident from Figure 3-2, the subject reacted by a moderate increase in heart rate at the opening incident when two masked and armed "robbers", one with a pistol and one with a short-barreled shotgun, came rushing through the main door shouting their orders to those who were currently present in the bank. This was an expected happening. All of the bank employees were ordered to lie flat on the floor, face down. This caused the heart rate of our subject to drop. Then, all of a sudden, one of the "robbers" discovered the heart rate logger mounted on the back of our subject. This was unexpected and caused a considerable amount of real, nonsimulated turmoil. This, naturally, was unexpected by our subject, who was ordered to stand for a closer examination. This resulted in a sharp rise in his heart rate, as was also the case when he unexpectedly was taken to the front door and thrown out of the building. He was then facing a crowd of spectators and the police who had surrounded the building. A police interrogation followed. This caused another sharp increase in heart rate in the innocent bank teller who was well aware of the fact that he was dealing with a simulated incident. His heart rate remained elevated for about an hour after the incident (Figure 3-2).

This simple experiment clearly shows that even when the subject is properly informed beforehand about an incident of this kind and is aware of the fact that

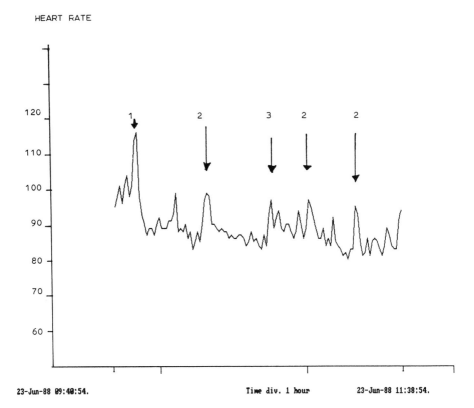

FIGURE 3-5. The heart rate recorded in a foreign exchange agent at the Oslo Stock Exchange during a trading session (1) Seated at his desk, getting ready to start trading; (2) walking; (3) trading the $75,000.

it is not real, and therefore harmless, it represents an amazingly great stress, as judged by the recorded heart rate reaction.

A similar recording, with the same sensor–logging combination, was done on a foreign exchange agent at the Oslo Stock Exchange during a trading session, when, according to the trader himself, there was little at stake as far as he himself was concerned. As is evident from Figure 3-5, his heart rate was highest during the initial, preparatory stage, probably due to apprehension, followed by a modest heart rate elevation when he traded $75,000 of shares on behalf of one of the companies he represented.

The same heart rate sensor–12-bit logger combination was used for the assessment of the physical work stress of some aluminum production plant operators in the north of Norway (Nes, Karstensen, and Rodahl, 1991b). Subsequently the purpose was to assess the effect of the ambient temperature on the heart rate in these operators (Nes, Karstensen, and Rodahl, 1991a).

In the first case, six operators (five men and one woman, ages 29 to 41) were subject to continuous ambulatory recording of the heart rate, combined with the recording of body and environmental temperature (Botsball temperature) during an entire work shift. In five of them, the recordings were made during two different shifts and repeated twice. Three of them were engaged in gas manifold changing, two in burner cleaning, and one in jack raising.

Most of the actual measurements of heart rate and other parameters were performed by members of the technical staff of the aluminum-plant laboratory, under the supervision of an experienced work physiologist. The heart rate was logged by a 12-bit Squirrel logger kept in a metal box in order to shield it from the powerful magnetic fields in the plant. Conmed pregelled disposable ECG electrodes (produced by Consolidated Medical Equipment, Inc., Utica, New York) were attached to the skin, as shown in Figure 3-3. The recorded data in the Squirrel memory were immediately transferred to an IBM-compatible PC in a room adjacent to the pot room and displayed on the PC screen in the form of graphs. This allowed the participating subjects and observers to view and discuss the results obtained and discuss their practical implications.

The study revealed, as could be expected, considerable individual differences in the work load of operators doing the same job. It also showed considerable variations from day to day in the same person, depending on how the electrolytic process was going. In the case of gas manifold changing, the study revealed in one of the male operators, brief periods of heart rate levels corresponding to a work load of about 50% of the operator's maximal work capacity. The female gas manifold changing operator, however, was exposed to work loads exceeding 50% of her maximal physical work capacity for periods of up to an hour and a half (Figure 3-6).

In jack raising, the physical work load approached 50% of the maximal work capacity during the 2 hours the jack raising operation lasted. In the burner cleaner, exceptionally high heart rates were observed. In one case, the average heart rate was 150 beats/min for as long as 60 min.

The study also revealed that during the actual work operations, the simultaneous logging of the Botsball temperature (as an index of heat stress) and the heart rate (as an index of work stress) may give a good indication of the relative effects of heat stress and work stress on the cardiovascular system. This was exhibited in operators exposed to both.

THE EFFECT OF HEAT STRESS ON HEART RATE

As was already mentioned, the effect of heat stress on heart rate was the subject of a subsequent study at the same aluminum-production plant. The study was done with workers engaged in heat-exposed jobs within a Soderberg pot room. It assessed the magnitude of the additional strain in terms of increased heart rate (Nes, Karstensen, and Rodahl, 1991a).

It is well known that heat stress may represent an additional load on the cardiovascular system. This is evidenced by an elevated heart rate at the same work load in a hot environment vs. a room-temperature environment. The explanation is that, in the case of heat stress, our circulating blood volume (in addition to having to transport oxygen) also has to serve as a cooling fluid. It, therefore,

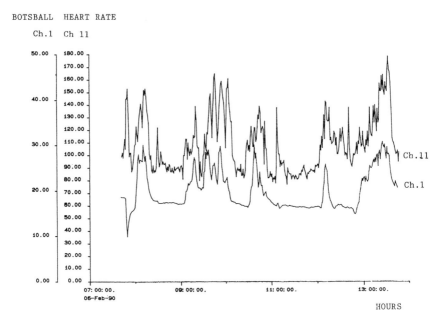

FIGURE 3-6. Relationship between ambient temperature (Botsball) and heart rate in a 34-year-old female operator in an aluminum production plant. (Ch. 1) Botsball temperature, in degrees Celsius; (Ch. 11) heart rate; (Ch = Channel). (From Nes et al., Elkem Aluminum Mosjøen Laboratorierapport, januar 1991a. With permission.)

transports heat from the interior of the body to the skin where it is dissipated to the surrounding environment by conduction, convection, radiation, and sweat evaporation. This requires an increase in speed of the blood circulation, i.e., the cardiac output (minute volume) has to be elevated. This can only be done by increasing the stroke volume of the heart and/or increasing the heart rate.

Since the possibility of increasing the stroke volume of the heart is limited, a major increase in the minute volume can only be achieved by an increase in the heart rate. Thus, the heart rate becomes an expression of the magnitude of the additional load exerted on the cardiovascular system when the body is exposed to a certain heat stress (for further details, see Åstrand and Rodahl, 1986).

It is, therefore, to be expected that the heart rate increases with increasing body temperature. In turn, this is affected by the temperature of the environment as well as the work rate. It is, therefore, not surprising that when recording the heart rate in workers operating in the Soderberg pot room, it was, in some cases, observed that the heart rate emulated the environmental temperature (Botsball Temperature Index). In other words, they fluctuated synchronously (Figure 3-6). This is in addition to the increased heart rate due to the physical work load.

The next step was to determine what part of the increased heart rate is caused by the heat stress in the workers operating in the pot room. In six subjects at rest and at two different work loads on a cycle ergometer at room temperature, a systematic registration of the heart rate was made in the laboratory (mean Botsball temperature 15.4 to 16.4°C) and at a selected location in front of a Soderberg pot in the pot room (mean Botsball temperature 22.0 to 23.1°C). Each subject was studied on two different days. The first day the subject was studied at normal room temperature in the laboratory in the morning. The subject rested in a sitting position for 15 min. This was followed by a 10-min cycle ergometer exercise at a load of 300 kpm/min, then a 5-min rest in a sitting position, and, finally, a 10-min cycle ergometer exercise at a load of 600 kpm/min followed by a 60-min rest at normal room temperature. The subject then went to a selected place in front of a pot in the Soderberg pot room, where he or she remained seated in a chair for 45 min in order for the body temperature to become adjusted to the prevailing ambient temperature. Following this, the program was identical to that followed at room temperature in the laboratory.

The second day, the same subject went through the program in reverse order, i.e., he or she started in the pot room and finished in the laboratory. This was done in order to adjust for possible effects of circadian rhythmic changes in body temperatures and heart rate as a consequence of the different time of the day.

The mean difference in the heart rate in the laboratory (room temperature) and in the pot room (heat stress) was as follows:

Location	Rest	300 kpm/min	600 kpm/min
Pot room	79	119	140
Laboratory	74	100	115
Difference	5	19	25

The results of this study show that the heat stress alone which faces the operator working at a rate of 300 and 600 kpm/min (50 and 100 watts/min) in front of a typical Soderberg pot, represents an additional load on the cardiovascular system in the order of 20 to 25 beats/min, corresponding to an approximate 20% increase in the work load. This should be taken into account in order to prevent undue fatigue in the workers engaged in jobs involving excessive heat exposure.

THE IMMEDIATE EFFECT OF SEVERE HEAT EXPOSURE ON THE HEART RATE

During a series of logging of the physiological effects of intensive heat exposure in a glass factory (Rodahl and Guthe, 1991; Rodahl et al., 1991a), it was observed that the heart rate reacted surprisingly quick on the ambient temperature, and, more or less, oscillated synchronously with the temperature surrounding the subject. This finding was of particular interest because the increased heart rate, in the long run, represents an increased load on the cardiovascular system. This very rapid increase in the heart rate was not caused by physical activity since the subject was standing quietly in front of the heat source. It can, therefore, only be caused by the heat stress. This poses the question as to the physiological mechanism behind this elevated heart rate. The answer to this would be essential in order to counteract the effect.

As was previously addressed, there are a number of indications that the head, and especially the face, plays a key role in the body's reaction to heat stress (see Åstrand and Rodahl, 1986; Rodahl, K., 1989). Riggs et al. (1981) have shown in laboratory experiments that cooling the face in physically active subjects caused a drop in heart rate without any changes in blood pressure or rectal temperature.

In connection with a study of divers and the so-called diving reflex, it has been shown that cooling the face causes a reduced heart rate (Kawakami et al., 1967; Hurwitz and Furedy, 1986). The exact location of the specific receptors in the face that elicit the heart rate effect, however, is not clear.

In mammals, the diving reflex is elicited by submerging the head. The sensory inputs are assumed to be caused by neural signals from the face, causing cessation of respiratory movements as well as vasoconstriction and reduced heart rate (Dykes, 1974). In man, both dry and wet facial cooling reduces the heart rate (Hurwitz and Furedy, 1986; Kawakami et al., 1967).

On the other hand, it is well founded that an increase in the overall body temperature causes an elevation of the heart rate. This was already discussed. It has not previously been shown what effect a very brief heating of the head or the body may have on the heart rate.

In a study of one of the subjects in the glass factory project referred to previously (Rodahl et al., 1991a), the subject's heart rate, as well as ambient and skin temperatures, were recorded in repeated 10-min periods, standing inactively for 1.5 min away from the production line of red hot wine bottles. Between each exposure period, the subject spent 5 min inside a cooled control room for the purpose of changing head gear, etc. The heart rate was recorded with the aid of the 8-bit Eltek Special Squirrel with a built-in heart rate counter (Rodahl et al., 1992a).

In the first case (Figure 3-7), he was exposed bareheaded, dressed in normal work clothing. In addition, the trunk was shielded by a reflective aluminum-foil apron as well as a reflective jacket. This was done in order to determine the influence of the trunk skin temperature on his heart rate. The study also included the effect of cooling the face around the mouth and nose with the aid of ice bags taped to the face. Another question was the effect of cooling the area around the ears, and finally, the effect of the inhalation of cooled air by using a rubber tube connected to a bag containing ice-cooled air (Rodahl et al., 1992b).

It was found that the heart rate in this subject, standing inactive in front of a heat source, fluctuated synchronously with the ambient temperature, i.e., the hotter the environment, the higher the heart rate (Figure 3-7). It should be noted that the changes in the heart rate were extremely rapid.

From Figure 3-8A and B, it appears that the heart rate increases in the heat-exposed subject even though the body is protected by reflective clothing which prevents a rise in the body skin temperature. However, if the protective clothing is removed (Figure 3-8C) the body skin temperature rises and a further increase in the heart rate is observed.

It appears from Figure 3-9 that cooling of the face around the mouth and nose, in addition to trunk shielding, does not reduce the elevated heart rate in the heat exposed subject.

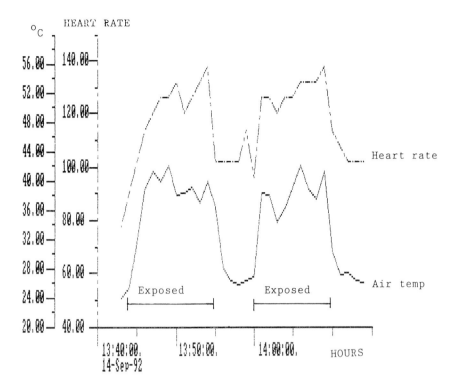

FIGURE 3-7. The heart rate increases synchronously with the air temperature in a subject exposed twice for 10 min each time in front of a glass bottle production line (1.5 m away), interspersed with a 5-min stay in a control room at normal room temperature. The subject is bareheaded, dressed in normal work clothing with an aluminum-foil reflective jacket and apron.

Cooling of the area around the ears did not seem to have any effect. Nor did the visor in front of the helmet and covered on the outside by aluminum foil seem to have an effect. The attempt to have the subject inhale cooled air while being exposed to the same heat source gave inconclusive results.

On the basis of these exploratory experiments which involved only one subject, one may feel encouraged to proceed with more systematic investigations of this problem area. However, one may perhaps be justified, already at this early stage, in assuming that the rapid increase in heart rate in subjects suddenly exposed to heat may be mediated through some sort of a trigger mechanism which causes the blood vessels in the skin to automatically be diluted. This reduces the circulatory resistance through the skin. Since the blood pressure is a product of the cardiac minute volume (i.e., stroke volume times heart rate) and the peripheral resistance, a reduction of the latter may cause a drop in the blood pressure. In order to prevent this, the minute volume of the heart has to increase. This can only be done by increasing the stroke volume and/or by increasing the heart rate. Since the possibility of increasing the cardiac stroke volume is limited, the result will essentially be an increase in the heart rate. This is, in fact, what we have observed.

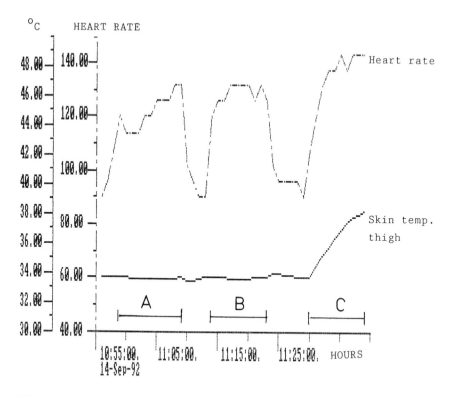

FIGURE 3-8. Heart rate and thigh skin temperature in a subject exposed to the heat of a glass bottle production line (1.5 m away), interspersed by a 5-min stay in a control room at normal room temperature: (A) bareheaded, dressed in work clothes and aluminum reflective jacket and apron; (B) dressed as (A), but now with an ice bag strapped in front of his mouth and nose; (C) bareheaded, dressed in work clothes only.

THE 8-BIT SQUIRREL WITH A BUILT-IN HEART RATE COUNTER: THE ELTEK SPECIAL

This 8-bit Squirrel is an ECG pulse generator in combination with 10 inputs for ambient and body temperatures. It weighs less than 700 g and can be carried in a leather case, in a belt, or by a strap over the shoulder.

Before the Eltek Special Squirrel is in recording mode, the counting period for the heart rate (10 to 60 seconds) should be set by using Function 7 and pressing Button B to select 7.1, and pressing Button C to select your choice, say 15 (i.e., a counting period of 15 seconds). The choice of channels is made by Function 9, using Buttom B to select Channel 12, and Button C to change the indication from 12.0 to 12.1. The timing for storing the reading in the memory is done by Function 6, using Button B to select minutes or seconds, and Button C to select the desired recording intervals. The correct time is set by Function 4, using Button B to select hour and minutes, and Button C to install the correct time. The date, in terms of

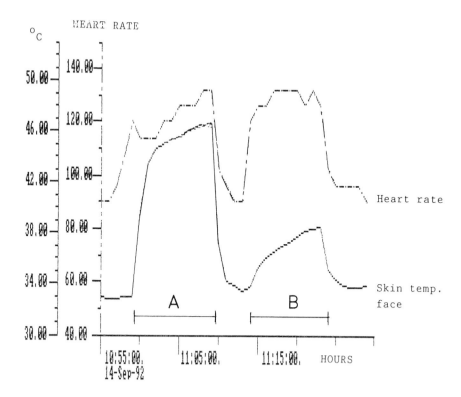

FIGURE 3-9. Heart rate and face skin temperature in a subject exposed bareheaded to the heat of a glass bottle production line (1.5 m away), interspersed with a 5-min break at normal room temperature: (A) dressed in normal work clothing and an aluminum reflective jacket and apron; (B) dressed as (A), but now having an ice bag strapped in front of his mouth and nose.

months and days, is set by Function 5. Function 1.1 is now used to start the recording by holding Button B down for a couple of seconds. The same button is used to stop the recording. Pressing Button C, in Function 1.1, and holding it for a few seconds will delete all data stored in the memory.

Examples of the use of this Eltek Special heart rate recorder in a subject in a glass factory are shown in Figures 3-7, 3-8, and 3-9.

In our experience, continuous recording of the heart rate permits an uninterrupted collection of data which reflects the subject's work load during the entire work day. A quantitative numerical analysis of the recorded data paired with detailed activity logs and supplemented by visual analysis of the replayed heart rate graphs, permits a comprehensive and dynamic evaluation of the circulatory strain imposed by work loads of varying intensity. The use of loggers and computers makes it possible to analyze large series of observations and ambulatory recordings of environmental stresses and physiological reactions to them. It also enables the observer to assess the relative severity of the work load, and the occurrence and duration of work loads exceeding a certain percentage of the individual's maximal aerobic power, or how much of the individual's heart rate reserve (the maximal heart rate minus the resting heart rate) is taxed in the performance of certain tasks or work operations.

For practical purposes, the heart rate, as such, may also be used to indicate the severity of the work load. A heart rate of 130 in most individuals roughly corresponds to about 50% of the individual's maximal aerobic power.

CHAPTER 4

Squirrel-Based Ambulatory Logging of Muscle Tension as an Expression of Muscular Work Load

A large number of individuals are unable to tolerate prolonged, intense, *static* muscular work as a regular occupation without ill effects. These effects include muscular stiffness, pain, tension, and even symptoms and signs of neuromuscular disorders, and represent a major cause of sick absenteeism in many industrialized countries. In a questionnaire survey, Tola et al. (1988) found that about half of the 1194 machine operators questioned had had neck and shoulder symptoms during the preceding 7 days, as against only 24% of the office workers. Westgaard et al. (1984), made a survey of musculoskeletal complaints in young seamstresses in four different factories. The absences due to musculoskeletal complaints was quite different in the different factories, although they were doing similar work. In these four factories, musculoskeletal or neuromuscular complaints accounted for between one quarter and one half of the total sick absenteeism, and this absence was not related to age. Within 3 years of employment, close to half of the seamstresses had been absent from work for prolonged periods because of musculoskeletal complaints (Figure 4-1).

In spite of considerable investments to improve the working conditions ergonomically in the above-mentioned clothing factories, no convincing evidence of reduced absenteeism due to musculoskeletal complaints was observed. In two of the factories, absence due to neuromuscular complaints increased, in one it remained unchanged, and in one it decreased (Westgaard et al., 1984). Prolonged intense static muscular tension or working posture may not be the only factors leading to musculoskeletal complaints. This was indicated by the findings of Tola et al. (1988) which showed that age and job satisfaction proved to be significant risk indicators for neck and shoulder symptoms. They also showed that working in twisted or bent postures may be a causative factor in the development of such symptoms. In this respect, it should be kept in mind that body posture, in general, may effect the relative load on different muscle groups. The center of gravity of

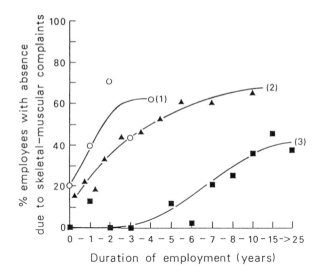

FIGURE 4-1. Percentage of employers with absences due to musculoskeletal complaints, in relation to duration of employment: (1) workers at an electronic assembly plant; (2) seamstresses; (3) office workers. (Modified from Westgaard et al., Arbeidsfysiologisk Institutt, Oslo, AFYI Publikasjon, 1984.)

the head and trunk is very close to the supporting column of bones. Because of this, the human being has the most economical antigravity mechanism among the mammals once the upright posture is attained. During forward flexion of the spinal column, there is a marked muscular activity, until flexion is extreme, at which point ligament structures take over the support of the trunk, and the electromyographically recorded discharge from the trunk muscles ceases (Floyd and Silver, 1955; Basmajian, 1979).

In the past, it has been generally assumed that the musculoskeletal and neuromuscular complaints previously described are the result of prolonged, static muscular tension. The fact is, however, that a review of the available literature reveals amazingly few studies of the relationship between objectively recorded muscle tension and musculoskeletal or neuromuscular symptoms based on actual observations or measurements on performing individuals in the field. Norman (1991) has briefly reviewed some of these problems as an occupational epidemic that goes by many names. Philipson et al. (1990), have shown that violinists who suffered from shoulder and neck problems had a higher mean load on some of the muscles involved in the dynamic muscular activity of playing the violin than did violinists without such problems. There was no difference in the level of muscular tension between the two groups at rest. To what extent some of the differences observed may be due to differences in maximal muscular strength is not clear from the presented data. Nor is it clear from this study to what extent static muscle tension is involved in the playing of a violin. The authors emphasize that the study was performed in the laboratory and that the equipment which was used for the study was relatively bulky and could not be used outside the laboratory. They

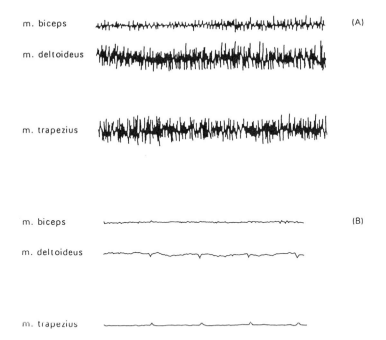

FIGURE 4-2. The difference at the end of 3 min in EMG signals from the biceps, deltoideus, and trapezius muscles when (A) the hand is firmly grasping the middle of the telephone receiver, holding it against the ear, while the upper arm and elbow are kept away from the chest at an angle of about 60°; and (B) the hand is holding the lower end of the telephone receiver, and the elbow is resting against the chest. (From Rodahl, K., 1989)

pointed out the need for the design and development of portable loggers capable of recording the electromyogram (EMG) in performers during real concert situations, since the EMG method does provide an objective measurement of occupational muscular load.

It is obvious that in order to determine the causative relationship between static muscle tension and neuromuscular complaints, objective data on the magnitude and duration of muscle tension is essential. This is true, not only in laboratory experiments, but also in individuals doing static muscular work in real life (Rodahl and Guthe, 1990).

The contraction of the fibers in a muscle is associated with small electrical changes, or oscillations, which can be detected by suitable surface electrodes placed on the skin over the muscle in question. The instrument picking up the electrical oscillations is known as an electromyograph, and the display of the myoelectrical oscillations is called an EMG. A simple example of such a direct electromyographic recording is presented in Figure 4-2. The greater the load on the muscle, the larger the oscillations. In more sophisticated quantitative EMG, the generated electrical activity is measured by rectifying and integrating the action potentials recorded from surface electrodes. These integrated action potentials are linearly or semilinearly

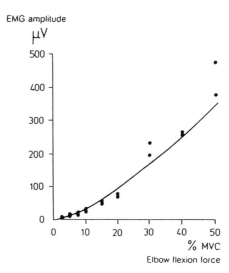

FIGURE 4-3. Relationship between EMG activity from the biceps brachii muscle and elbow flexion force as a percentage of the maximal voluntary contraction (% MVC). (From Hagberg, M., Vetenskapelig Skriftserie, Arbetarskyddsverket, Stockholm, Sweden, 1981. With permission.)

related to the muscular force which is being exerted (Bigland and Lippold, 1954) (Figure 4-3).

Integrated electromyographic recordings will permit the measurement of the load on single muscles, as well as groups of muscles. It is well established that the EMG does reflect the magnitude of the muscle engagement and may be used to measure the exerted forces in percent of the maximal voluntary muscle strength. For further references, see Rodahl, K. (1989).

This can be done by a pocket-sized, battery-operated EMG, such as the Squirrel- compatible Myolog, produced by Aleph One, Ltd., Cambridge, England. It uses three disposable skin electrodes to detect the signals from the muscle in question (Figure 4-4). The Myolog is plugged into the Squirrel (Channel 5, 6, 7, or 8 on the 1201 series Squirrel) and both instruments are attached to a belt and carried by the subject. This combination with a Squirrel provides an ambulatory system which can be used for the purpose of logging the tension in certain muscle groups over extended periods, even 24 hours, to find out when and why excessive tension develops in the muscle groups which are being monitored (Figures 4-5, 4-6, and 4-7).

In order to relate the amplitude of the EMG signals to known levels of muscular tension, a routine calibration procedure may be introduced at the beginning and again at the end of the recording period. This is done after the electrodes are in place and the logging system has been started. This can be done when the shoulder muscles are being recorded by first asking the subject to stand upright in a relaxed position with both arms hanging passively alongside his body for a period of some 15 seconds. This may then be followed by asking the subject to raise his arms in

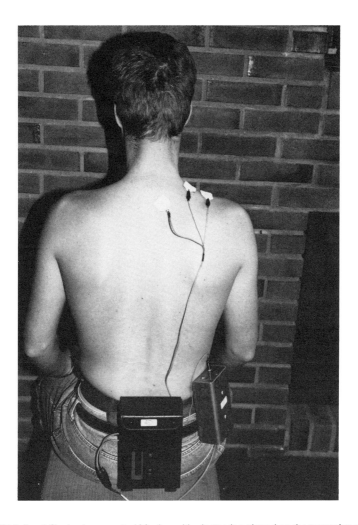

FIGURE 4-4. A Squirrel-connected Myolog with electrodes placed on the trapezius muscle.

a horizontal position and hold this position for some 15 seconds, followed by a new period of relaxation (Veiersted, 1991). This may be supplemented by placing known weights at the subject's elbow or wrist while the muscle tension is being recorded, or by placing known weights in the subject's hands. An example of this is shown in Figure 4-8. In the case of recording the muscle tension in the lower arm muscles, the subject may be asked to grasp a tennis ball in his hand and to squeeze it as hard as he can. For more scientific purposes, it is possible to arrange for a procedure involving the recording of the maximal voluntary force which the subject can attain in the muscle in question. On this basis, the recorded level of muscle engagement may be expressed in percentage of the maximal voluntary force (%MVF).

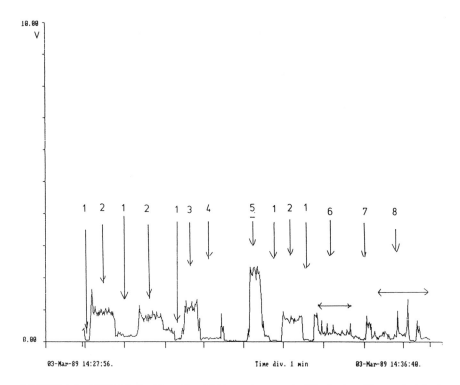

FIGURE 4-5. EMG (Myolog) from the right trapezius muscle, (1) relaxing; (2) arm stretched out horizontally; (3) holding telephone receiver, elbow extended; (4) holding telephone receiver, elbow supported; (5) lifting chair; (6) writing; (7) standing; (8) walking.

The level of EMG amplitude may vary greatly from person to person, even when the electrodes are placed over the same muscles as identically as possible. The EMG amplitudes may also vary considerably from time to time in the same subject, depending on the exact location of the electrodes (Veiersted, 1991).

For comparative recordings, therefore, it is advisable to leave the electrodes in place and to make the recordings from the same electrodes during all the experimental conditions which are being compared.

THE USE OF THE MYOLOG–SQUIRREL COMBINATION FOR THE LOGGING OF MUSCLE ENGAGEMENT

A major advantage with this ambulatory EMG logging system is that it can be used to visualize the consequences of the practical working procedures on the load imposed on the muscles involved. This is evident from a number of studies recently made in modern industrial operations, as well as office work, including the use of personal computers with and without the use of a Mouse. Such a

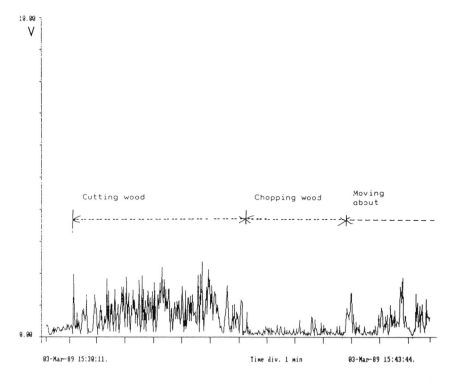

FIGURE 4-6. EMG (Myolog) from the right trapezius muscle, recorded once every second.

recording of reality may then be used as a basis for the implementation of improvements. A few selected examples will be presented.

During part of the work shift, a series of EMG recordings were made from the trapezius muscle in workers engaged in the production of plastic bottles and plastic containers. In one of the factories, the subjects were engaged in the printing of different labels on plastic bottles, and the packing of the finished product in large cardboard boxes (Rodahl and Guthe, 1990). The EMG tracings from two of the operators, with comparable voltage and time axes, are shown in Figures 4-9 and 4-10.

The work of one of the operators consisted of emptying plastic bottles from large cardboard boxes into a container from where the bottles automatically were picked up by a conveyor belt and taken through a printing device where colored labels were printed on the bottles. This operation was carried out by two women who alternated their jobs every 30 min. In addition to the above-mentioned procedures, they removed the filled cardboard boxes at certain intervals with the aid of a trolley. A 10-min break was included in the observation period.

From Figure 4-9 it is evident that the emptying of plastic bottles from the large cardboard boxes into the container that feeds the conveyor belt represented typical dynamic work with brief working periods. From a physiological point of view, this operation appeared by no means unreasonable. The other part of the job,

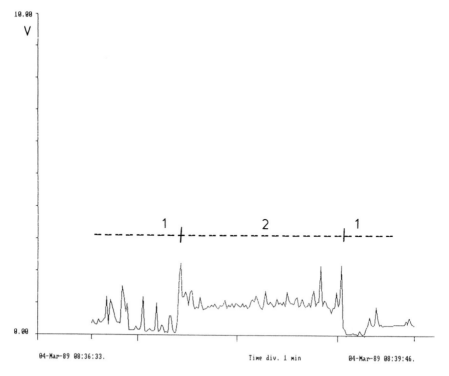

FIGURE 4-7. This EMG recording (Myolog) was made with electrodes which had been in place over the trapezius muscle for 24 hours: (1) moving; (2) reading the newspaper, sitting.

dealing with the printing operation, was on the whole more relaxed. The muscular work load was not much greater than the muscle tension recorded during the breaks. Generally speaking, the entire work operation appeared to be good, with a high degree of dynamic variation and opportunities for the operators themselves to influence the work pace and thus determine the work load.

The other subject (Figure 4-10) was engaged in the printing of labels on plastic ketchup bottles. Every 20 min, she changed from the job of checking the finished bottles at the end of the conveyor belt and packing them into cardboard boxes at the end of the line, to the job of feeding the bottles onto the conveyor belt. Figure 4-10 shows a significantly greater muscle tension in the right-shoulder muscles during the checking operation than during the feeding of the bottles onto the conveyor belt. In both cases, however, it is a persistent, static muscle tension, greatest during the checking operation. This was due to the fact that the subject failed to rest her arm when not in use during the work. Only when the conveyor belt was stopped (for the purpose of cleaning and oiling some of the moving parts), did the subject rest the arm in her lap. This decreased the muscle tension to the resting level. It should also be mentioned that during all the operations involved she only used the right hand and rested the other, instead of alternating, i.e., using one hand at a time while resting the other.

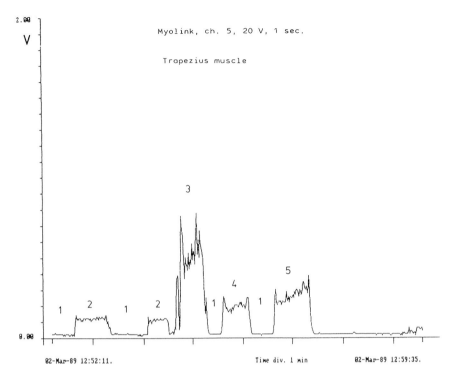

FIGURE 4-8. EMG calibration (the right trapezius muscle): (1) relaxing; (2) arm stretched out horizontally; (3) lifting 10 kg; (4) lifting 2 kg placed at elbow; (5) lifting 2 kg placed at wrist.

A further series of three examples is taken from a study at a plastic container factory (Rodahl et al., 1991d). The EMG for the right trapezius muscle was logged for three subjects. The first subject, a short 22-year-old girl, was engaged in the packing of finished plastic containers at the end of a conveyor belt serving several production units. The containers had to be removed from the assembly line and stacked on top of each other. This, in the end, became quite difficult for the short girl, who then had to climb on a ladder in order to reach the top of the pile. As is evident from Figure 4-11, the subject is able to relax her muscles completely, both when asked to do so and when it is possible to relax during work. During the early phases of the observation period, complete relaxation does occur at brief intervals with tension levels down to zero (phase A in Figure 4-11). During the last 25 min of the observation period, however, the tempo is markedly increased because of additional production units being activated. This causes significantly higher EMG amplitudes and persistent muscle tension without relaxation (phase B in Figure 4-11). As judged by the reaction of the subject, it appeared that this final work rate in the long run probably would have been too much for this subject. In this case, an additional recording of her heart rate as an expression of the physical work rate would have provided additional valuable information for the assessment of a tolerable work load.

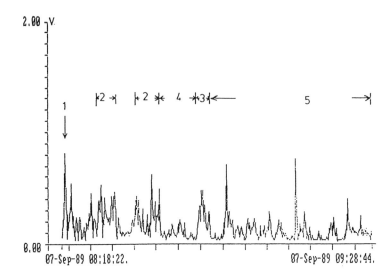

FIGURE 4-9. EMG amplitudes (trapezius muscle) in a female worker in a plastic bottle factory: (1) calibration; (2) transferring bottles into container; (3) moving containers; (4) break; (5) printing labels on the plastic bottles. (From Rodahl, K. and T. Guthe, Arbete mænniska miljö, 2.90:116, Stockholm, 1990. With permission.)

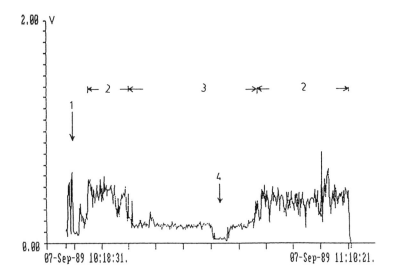

FIGURE 4-10. EMG recordings of the right trapezius muscle in a female operator engaged in the printing of labels in plastic ketchup bottles: (1) calibration; (2) control of the finished bottles, sitting; (3) feeding bottles on to the conveyor belt, sitting; (4) the conveyor belt is stopped, and the subject is resting her arms in her lap, sitting. (From Rodahl, K. and T. Guthe, Arbete mænniska miljö, 2.90:116, Stockholm, 1990. With permission.)

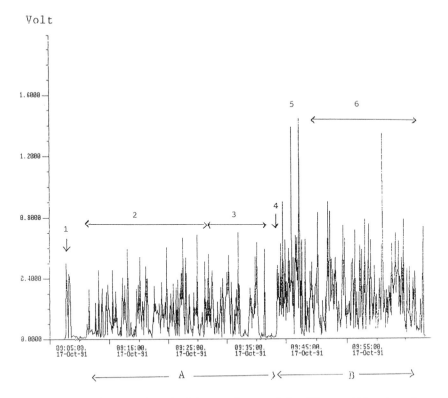

FIGURE 4-11. EMG logging from the right trapezius muscle in a 22-year-old female worker, packing plastic containers at the end of a conveyor belt: (A) moderate tempo; (B) increased tempo; (1) calibration; (2) removing containers from the conveyor belt; (3) stacking containers; (4) moving stacks of containers; (5) climbing a ladder; (6) additional production units activated. (From Rodahl et al., KIL-amil-dok-9, 12 oktober, 1991d. With permission.)

The second subject was a 22-year-old, strong young man involved in the production of plastic gasoline tanks for automobiles. From Figure 4-12, it can be observed that he works almost continuously at an even work rate. His work is dynamic and varied, and his EMG amplitudes during work are below the level attained during the calibration when he was asked to raise his arms to a horizontal position at the level of his shoulders [(1) in Figure 4-12]. Thus, the work itself represents a rather moderate work load for this individual. On the other hand, Figure 4-12 reveals a persistent elevated muscle tension without any proper relaxation during the entire observation period, compared to the level of relaxation exhibited during the calibration [(2) in Figure 4-12]. This persistent muscular tension may possibly be a sign of mental or psychomotor tension.

The third subject was a 21-year-old male operator. His work consisted of a highly routine operation in a standing position. He lifted the finished plastic containers from the machine where the containers were being produced. Once every other minute or so, he removed excess plastic around the opening and placed

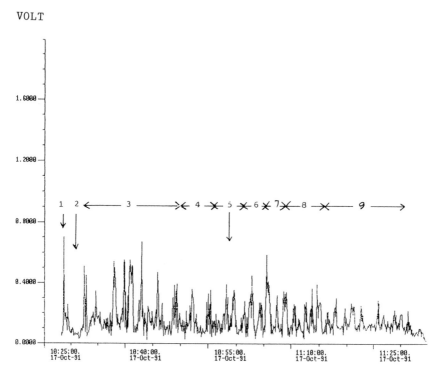

FIGURE 4-12. EMG logging from the right trapezius muscle in a 22-year-old worker involved in the production of plastic gasoline tanks for automobiles: (1) calibration; (2) relaxing; (3) routine treatment of the produced tanks; (4) preparing containers; (5) cutting up faulty tanks: (6) same as (4); (7) same as (3); (8) same as (4); (9) same as (3). (From Rodahl et al., KIL-amil-dok-9, 12 oktober, 1991d. With permission.)

the containers in an orderly manner on a table for cooling. When cooled, they were stacked on a platform on the floor. Occasionally, he would take a break, standing, while waiting between the different operating routines. As is evident from Figure 4-13, this operator is subject mainly to prolonged periods of static muscular work loads interspaced with some incidents of muscular relaxation with EMG amplitudes down to zero during some of the waiting periods. On the basis of these EMG recordings, it may be concluded that the subject is able to relax his muscles completely when resting. The prolonged periods of persistently increased muscle tension, therefore, must be caused by the work which to a large extent is static in nature. Furthermore, the work is done with his arms and shoulders in an elevated position. From a physiological point of view, this is a rather undesirable work operation which should be replaced by machines capable of performing most, if not all, of the work operations involved.

A third set of examples was taken from a study of some of the operators at the assembly line in a floorboard production company (Figures 4-14 and 4-15). This study compared two female operators sitting next to each other and doing the same

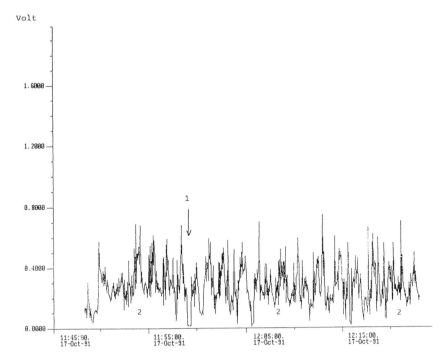

FIGURE 4-13. EMG logging from the right trapezius muscle in a 21-year-old male operator, removing plastic containers from a container-producing machine in a standing position: (1) waiting; (2) persistent, static muscle tension. (From Rodahl et al., KIL-amil-dok-9, 12 oktober, 1991d. With permission.)

sorting of pieces of wood on the conveyor belt. One of them used every opportunity to rest her arms at the edge of the table, thus creating a dynamic type of muscular work (Figure 4-14), while the other subject kept her arms unsupported in front of her while waiting for the next pieces of wood to appear on the conveyor belt (Figure 4-15) (Rodahl, K., 1989).

The same EMG logging procedure was used in a study of the muscle tension of the shoulder muscles in three operators grinding blocks of crystal glass in a glass factory (Guthe et al., 1990). This study used the combination of a Myolog EMG-recorder and a 1202-series 12-bit Squirrel logger.

In the first case, the subject was grinding a large slab of crystal glass in a standing position, leaning forward, while both elbows rested on pillows. This was done in order to use the weight of his body to increase the pressure against the rotating grinding disk. He held the glass slab with both hands and turned it with the help of his thumbs. From Figure 4-16, it can be seen that the shoulder muscles are activated rhythmically with brief contractions, separated by brief periods of relaxation. The lifting of a basket containing heavy glass slabs causes very high muscle contractions, as could be expected.

The next subject (Figure 4-17) was grinding a particular pattern of figures into heavy crystal wine glasses. The subject held the glasses horizontally and pressed

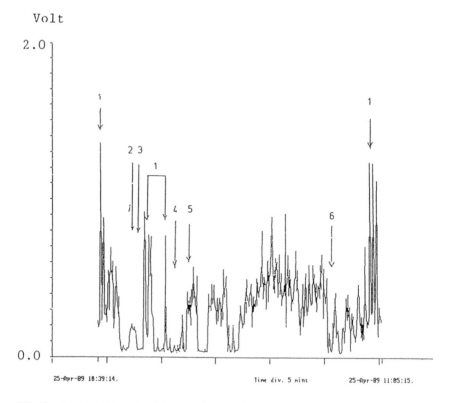

FIGURE 4-14. EMG from the right trapezius muscle in a female operator at the assembly line in a floorboard production plant, resting her arms effectively at the edge of the table between each work operation: (1) calibration; (2) starting to work; (3) waiting; (4) relaxing; (5) starting to work again; (6) brief pause in the work. (From Rodahl, A., 1989)

them against the rotating cutting disk. This was usually done in a sitting position, but in this case, because the Squirrel logger interfered with his normal sitting posture, he changed to a standing position. As is evident from Figure 4-17, the static load on his shoulders was greater when he was working in a sitting position than in a standing position. This is probably because he had to press harder with his shoulders when sitting than when standing, because he was free to use more of his body weight to exert pressure against the cutting disk when he was standing. Although this work, basically, was rhythmic in nature, the working position and the degree of precision required may tend to make this type of work fairly static in nature.

The third subject (Figure 4-18) was grinding a large thick glass plate, which weighed about 6 kg. The final product was to look like an ice floe and to serve as a foundation for a large polar bear made of glass. He started by wet-grinding both sides of the plate and pressing it against a large, horizontally rotating, grinding disk (see 1 in Figure 4-18). He then carried it to another grinding machine where he ground the edges of the "floe" by pressing it against the grinding disk;

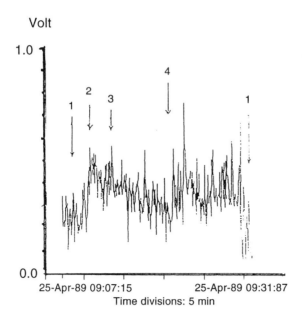

FIGURE 4-15. EMG from the right trapezius muscle in another female operator involved in exactly the same work operation as the subject in Figure 4-14. In this case, the operator does not rest her arms between each work operation, as was the case in Figure 4-14. (1) Calibration, (2) starting to work, (3) talking while working, (4) putting on ear protectors. (Rodahl, A., 1989)

his elbows were flexed about 90°. As is evident from Figure 4-18, the first operation did not pose much of a load on his shoulders. The following operation, on the other hand, involved the edge cutting of the heavy glass slab which he held in his hands, without any support. This represented a major static load on his shoulder muscles. It appears from Figure 4-18 that his trapezius muscles were contracted more or less the entire 35-min duration of this operation. From the results of this study, it is obvious that there must be better and less stressful ways of grinding the edges of a large glass slab.

AMBULATORY LOGGING IN THE ASSESSMENT OF THE CLAIMED RELATIONSHIP BETWEEN MUSCLE TENSION AND NEUROMUSCULAR COMPLAINTS

As already mentioned, it has long been taken for granted that prolonged, static muscular tension is the cause of neuromuscular or musculoskeletal complaints in the neck, shoulder, and arm regions. It has been assumed that the increased muscle tension is either caused by the working posture, the way the work is performed, or by an increased level of mental or psychomotor strain. These assumptions are largely based on clinical experience since the necessary tools have not been available in the past to objectively record the level of EMG tension in the muscles

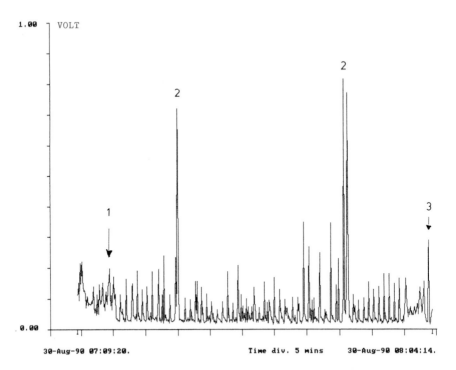

FIGURE 4-16. EMG logging of the right trapezius muscle in an operator grinding blocks of crystal glass in a glass factory: (1) start of work; (2) lifting heavy baskets; (3) calibration. (From Guthe et al., KIL-amil-dok-4, 25 Sept 1990. With permission.)

while the patient is carrying out the work in question, especially in the case of nonstationary occupations. This also applies to the recording of central nervous system–related muscle tension, e.g., in the case of the stress affecting the patient when he is off work. The type of instruments that record the EMG have so far been available for stationary but have not been very suitable for ambulatory logging in the field.

Now this is possible, however, with the ambulatory, battery-operated minicomputer system based on the Myolog EMG recorder combined with the ambulatory Squirrel logger, which not only displays the recorded values, but also stores them in the electronic memory for immediate display and printout in the form of graphs, etc. This EMG–logging system has provided a practical opportunity to relate the subjective symptoms of the usual types of neuromuscular complaints to EMG-recorded neuromuscular tension in the muscle groups in question. This can be done during rest and work and, in particular, during the type of muscular engagement thought to be the cause of the trouble.

An exploratory study was made in 16 employees in an insurance company, who for the last 5 to 20 years had been engaged in normal office work. Of the 16 employees, 12 had suffered from neck and shoulder complaints during the preceding 12 months (Maltun, 1990). In some of the cases, the complaints were severe

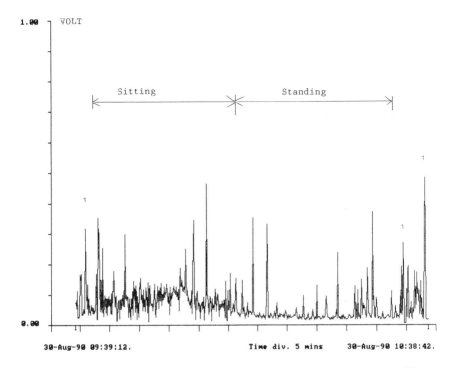

FIGURE 4-17. EMG from the right trapezius muscle in a subject grinding a pattern of figures into heavy crystal wine glasses, in a sitting and in a standing position: (1) calibration. (From Guthe et al., KIL-amil-dok-4, 25 Sept 1990. With permission.)

enough to make it necessary for the subjects to change jobs or prevented them from performing their ordinary daily work 1 to 7 days per year.

In some of these subjects with symptoms of neuromuscular complaints, the study revealed some degree of prolonged muscle tension and inadequate muscle relaxation during work (Figure 4-19), but others did not. On the other hand, one of the subjects free of neuromuscular complaints had some degree of intermittent static muscle tension, while the others did not (Figure 4-20). Although this was only a pilot study, with observation periods of no more than 60 min or so, it did, in a preliminary manner, question the assumed correlation between static muscle tension and neuromuscular complaints in office workers. All of the subjects examined, whether they had complaints or not, were able to relax completely during rest or when asked to do so. This indicated no evidence of enhanced psychomotor tension.

When palpating the neck and shoulder muscles, practicing physiotherapists may detect patients suffering from a stiff, sore neck. They may feel the muscles are tense, taut, and string-like in consistency (Bottolfsen, 1991). It has been assumed that this may be a sign of increased muscle tension. It is also generally believed that acupuncture may be an effective treatment in some of these cases. In the available literature, however, there is a paucity of published information on

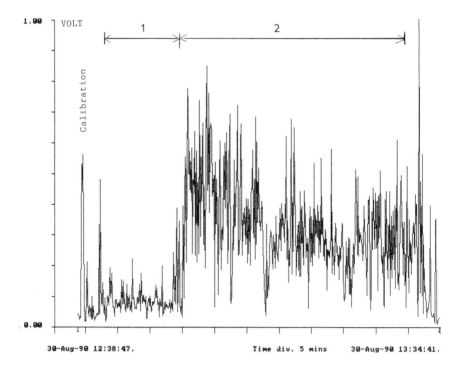

FIGURE 4-18. EMG from the right trapezius muscle in a subject grinding a large 6-kg glass plate into the shape of an ice floe: (1) wet-grinding both sides of the plate; (2) grinding the edges of the plate. (From Guthe et al., KIL-amil-dok-4, 25 Sept 1990. With permission.)

the effect of acupuncture on muscle tension, assessed by objective EMG measurements. Cunxin et al. (1989), in a study of clinical therapeutic effects of treating *soft tissue injuries* by acupuncture, did observe a significant difference in the EMG amplitude on the affected side of the lumbar area, before and after acupuncture treatment.

During one acupuncture session, Bottolfsen (1991) recorded the EMG amplitude from both the left and right trapezius muscles in a series of eight patients suffering from chronic complaints of pain and stiffness in the shoulder–neck region. All of the eight patients went through one session of EMG recording and acupuncture stimulation of the trapezius region. Five underwent a further session with the EMG electrodes placed on the trapezius muscles while the acupuncture points were on the leg. This was done in order to determine if the EMG amplitudes obtained differed according to which acupuncture points were used. The EMG was recorded from the left and the right trapezius muscles simultaneously: two surface electrodes were placed on the medial aspect of the left and the right trapezius muscle, the third neutral electrode was placed on either side, near the spine. The electrodes were connected to the Myolog EMG recorder (Aleph One Ltd., Cambridge, England), where the EMG signals were rectified and integrated in the usual manner. The signals from the Myolog were continuously transferred

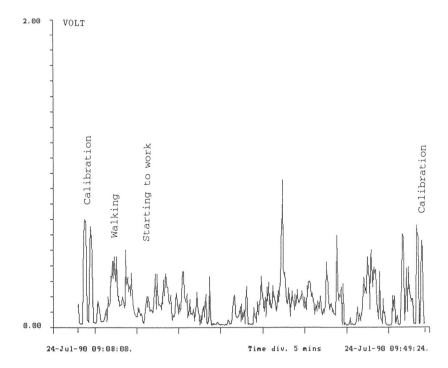

FIGURE 4-19. EMG from the right trapezius muscle in an office worker who had symptoms of neuromuscular complaints, showing some degree of prolonged muscle tension, and inadequate muscle relaxation during work. (From Maltun, K. R., personal communication, 1990.)

to a Squirrel multichannel electronic minilogger (Grants Instruments, Ltd.) which was carried on the person.

As a calibration procedure, the EMG was recorded while the subject was standing and relaxed, with his/her arms hanging alongside the body. The arms were then raised to a horizontal position at shoulder height, abducted 90° and forward flexed 10° and held for about 15 seconds. The arms were then dropped and hung relaxed alongside the body (Veiersted, 1991).

Following the calibration procedure, the patient was asked to lie face down on the couch, comfortably relaxed, and covered with a blanket. The same standard procedure was followed for all patients, including the start of the logging, the application of needles, the time for manipulating the needles (10 and 20 min from the start), and the time for removing the needles (30 min after the needles were applied).

The results showed that the EMG tracings remained even and relaxed in almost all cases, except for the brief spiky elevation in the EMG when the needle was inserted and manipulated (Figure 4-21). As judged by the EMG tracings, there was no obvious effect of acupuncture on the EMG amplitude level (apart from the above-mentioned spikes during the insertion and manipulation of the needle) or on the EMG recorded muscle tension in either of the trapezius muscles, regardless

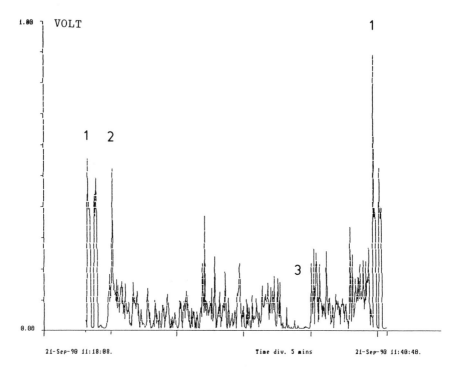

FIGURE 4-20. EMG from the right trapezius muscle in an office worker who did not have any symptoms of neuromuscular complaints, showing dynamic muscle contractions, interspersed with complete muscle relaxation: (1) calibration; (2) starting to work; (3) break. (From Maltun, K. R., personal communication, 1990.)

if it was applied to the upper or the lower extremity. In some cases, the EMG level appeared to be slightly elevated before acupuncture. This was most likely due to some degree of apprehension or anxiety before treatment. In the cases where upper extremity acupuncture was used, the EMG amplitudes were higher in the left trapezius, where the needle was inserted, than in the right trapezius. The insertion of the needle in the left shoulder, however, also caused increased muscle tension in the right shoulder.

Statistical analysis of the data from the eight patients who were subject to upper extremity acupuncture showed no statistically significant effects of the acupuncture on muscle tension (apart from the previously mentioned spikes) during the 30-min observation period. This was the case when the left and the right trapezius muscles were analyzed separately and when the data from both sides were combined. Both muscles reacted alike, although the tension of the left muscle remained slightly higher than that of the right. This difference was greatest 1 min after acupuncture. At zero time, before acupuncture, the difference was significant at a level of $p < 0.01$. At 1 min after acupuncture, it was significant at a level of $p < 0.001$. After 15 min, the level of significance had decreased to $p < 0.05$ and after 30 min, the difference was no longer statistically significant using the Student's T-test for paired data.

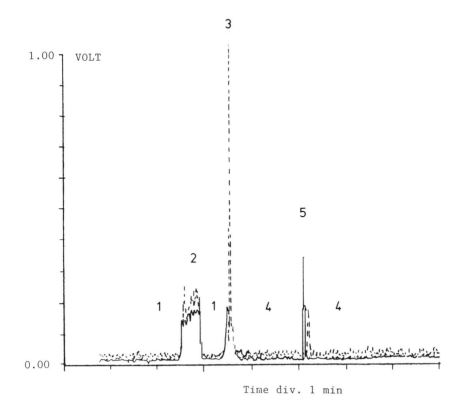

FIGURE 4-21. EMG tracings of the left (– –) and the right (—) trapezius muscles in an acupuncture patient: (1) standing relaxed; (2) arms elevated; (3) getting on couch; (4) lying relaxed on couch; (5) acupuncture. (From Bottolfsen, T., Thesis for acupuncture qualifying exam, Kristiansand, 1991. With permission.)

In the five patients who were subject to lower extremity acupuncture, there was no statistically significant difference in muscle tension between the right and the left trapezius muscle. Nor was there any statistically significant difference in the tension in either of the trapezius muscles from the beginning to the end of the observation period.

The purpose of the study was to find out whether or not a single acupuncture treatment of the upper and the lower extremity affected the level of EMG-recorded muscle tension of the trapezius muscle. The results did not reveal any such effect. Although this finding only applies to the conditions prevailing under this particular study, it may indicate that the mechanism for the observed acupuncture effect may be sought in parameters other than muscle tension, including the release of neuropeptides, psychophysiological functions, etc.

The fact that the EMG-recorded muscle tension in the left trapezius muscle was higher than in the right trapezius muscle in patients subject to upper extremity acupuncture may most likely be explained as a sign of nervous tension due to apprehension prior to the acupuncture procedure. This was known by the patient to take place on the left shoulder. This is supported by the fact that the state of tension declined in the course of the 30-min rest following acupuncture and by the fact that this difference was not observed in patients subject to lower extremity acupuncture.

With the introduction of extensive Mouse-based personal computer (PC) programs in the production of geological maps, reports, etc., at the headquarters of one of the major off-shore oil-producing companies in Norway, there was a noticeable increase in shoulder and arm complaints among the operators. This gave us the opportunity to study the effect of the use of the PC and the Mouse on the state of tension in the arm and shoulder muscles and to study to what extent such tension might be related to the arm and shoulder complaints. This was of particular interest since only one published report on the pathophysiological consequence of using the Mouse was found at the time. This report was a brief description by Norman (1991) of two cases of a condition which he names the "Mouse joint" with aching pain at the base of the second and and third fingers. He interpreted this as being the result of repetitive flexion activity of the index and middle fingers as required for Mouse operation.

Our field study, with systematic individual recording of muscle tension as evidenced by EMG-amplitude loggings, took place in February 1992 at the individual subjects' work place. The subjects were doing their normal daily work (Rodahl et al., 1992d). The study included five subjects who had symptoms and three who did not.

In all the subjects examined, whether they had symptoms or not, merely grasping or holding the Mouse in the hand while resting the hand or the arm on the table, without moving it and without pressing the button, caused a significant increase in the EMG amplitude recorded from the extensor digitorum muscle, as is evident from Figure 4-22 (subject without symptoms) and Figure 4-23 (subject with symptoms). The subjects, whether they had symptoms or not, were able to relax the examined muscles when asked to do so. It seemed to make very little difference in terms of EMG amplitude whether or not the subject supported the arm while using the Mouse (Figure 4-23). On the other hand, it was possible to alternate the use of the right and the left hand when using the Mouse, thus relieving the tension on the one arm (Figure 4-24).

Figure 4-25 shows the EMG amplitude of the right shoulder muscles in a 37-year-old woman who had just completed a 5-month sick leave due to intense aches and pains in her back, shoulders, arm, and wrist, considered to be related to her work involving the use of the Mouse. Figure 4-25 is a recording of the EMG from her right trapezius muscle, while working in her usual manner. It is quite evident that she has the ability to relax her muscles at will. Using both hands, touch-writing on the PC without arm support causes an elevated EMG amplitude level. The muscle tension in her right shoulder was significantly elevated when using the Mouse while resting her wrist on the table, but reduced to approximately resting level when she rested the entire arm on the table and continued to operate the Mouse. In Figure 4-26, the EMG is recorded simultaneously in the right extensor digitorum muscle and the right trapezius muscle in the same subject. Also, in this case, the tension level in the shoulder muscle is considerably less

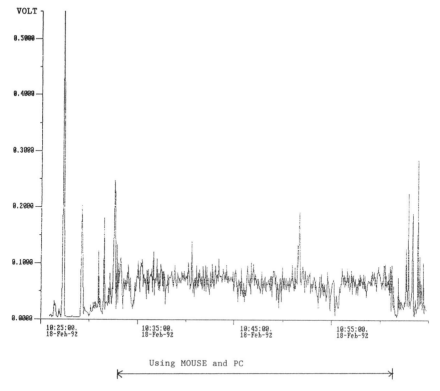

FIGURE 4-22. EMG recorded from the right extensor digitorum muscle in a PC operator using a Mouse with her right hand. This subject has no symptoms of neuromuscular complaints in spite of the persistent, static muscle tension. (From Rodahl et al., Project rapport, Norsk Hydro, Oslo, 1992d. With permission.)

when resting the arm on the table while using the Mouse. The resting of the arm has no effect, however, on the tension level of the extensor digitorum muscles, suggesting that the tension in the arm muscles is due to the grasping of the Mouse.

It is interesting that the recording of the state of tension in the shoulder muscles in the same subject (who suffered from aches and pains in her shoulder) when she was the key subject in a television recording (doing her usual work with the PC–Mouse combination and appearing to be visibly tense and nervous) revealed no evidence of elevated static muscle tension in her shoulder (Figure 4-27).

These observations clearly indicate that the use of a Mouse causes considerable static tension in the muscles of the arm using the Mouse. This is the case both in individuals who have and who do not have pains and aches in their arm muscles. It is, therefore, not obvious that static muscle tension associated with the use of Mouse-driven PC systems inevitably leads to neuromuscular disorders. The fact is that some individuals, even after several years of regular use of Mouse-driven PC systems, remain free of any symptoms of neuromuscular disorders.

There can be no doubt, however, that prolonged, monotonous static muscle tension does not represent a physiological muscle engagement and should be

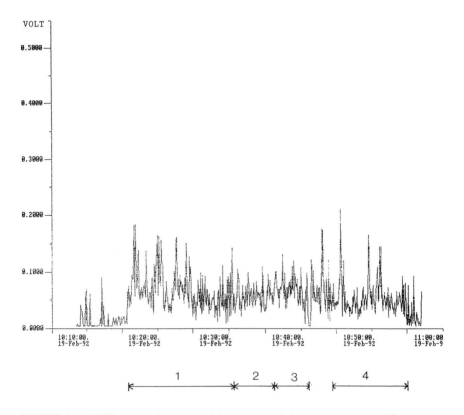

FIGURE 4-23. EMG recorded from the right extensor digitorum muscle in a PC operator using a Mouse with her right hand. This operator has symptoms of neuromuscular complaints: (1) using PC and Mouse without arm support; (2) using PC and Mouse while resting the distal end of the arm; (3) moving the Mouse with the first finger instead of the thumb; (4) using movable arm support while using the Mouse. (From Rodahl et al., Project rapport, Norsk Hydro, Oslo, 1992d. With permission.)

avoided because the human body, from a physiological point of view, is made for dynamic muscle activity, alternating between muscle contraction and relaxation.

Ambulatory logging of muscle tension may be a useful tool in the study of causative relationships between static muscular work loads and neuromuscular complaints. It may also be a useful instrument in the establishment of varied work routine, including intervals between static and dynamic work. Such ambulatory monitoring would also be useful in finding why some individuals do, and others do not, develop neuromuscular complaints when engaged in the same occupation.

This type of logging may also be used to test the ideal shape of the Mouse, causing a minimum negative effect. It may also be used to visualize the effect of transferring some of the Mouse operations to standard keyboard functions, in order to provide a more dynamic work situation. This may be of particular importance since it is essentially the hand grip of the Mouse which causes the static tension of the arm muscles. For this reason, the alternating use of the right

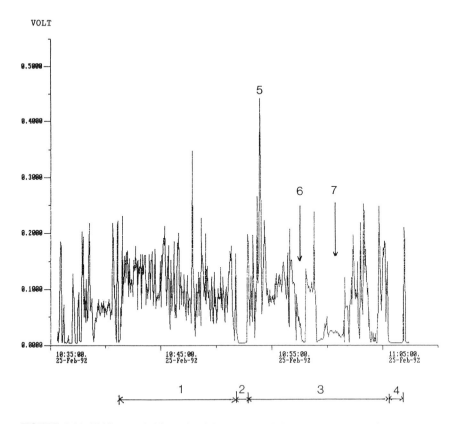

FIGURE 4-24. EMG recorded from the right extensor digitorum muscle in a PC operator (1) using a Mouse with the right hand; (2) relaxing both hands in the lap; (3) using the Mouse with the left hand; (4) relaxing, as in (2); (5) changing the position of the Mouse; (6) during a telephone call; and (7) using the PC with both hands. (From Rodahl et al., Project rapport, Norsk Hydro, Oslo, 1992d. With permission.)

and the left hand operating the Mouse, is also a possibility worth considering. Above all, a variation of the muscle engagement is essential. This includes occasional relaxation exercises to encourage the development of the ability to sense when the involved muscles are tensed and need to be relaxed.

Since the majority of the operators engaged in prolonged static work sooner or later may develop symptoms of neuromuscular disorders, and since there is no way of knowing in advance who is susceptible to developing such complaints, it would appear sensible, during orientation, to make all beginners aware of the dangers involved in prolonged, static work, and to give them instructions as to how to minimize the risk of developing such complaints. Furthermore, available evidence suggests that the development of symptoms of neuromuscular complaints may also be affected by factors other than static muscle tension, such as the social atmosphere, human relations, and interpersonal communication at the place of work, as well as other psychosocial factors. It appears that those who

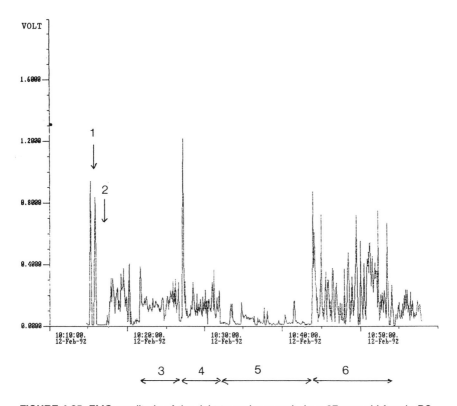

FIGURE 4-25. EMG amplitude of the right trapezius muscle in a 37-year-old female PC–Mouse operator who had just completed a 5-month sick leave due to intense aches and pains in her back, shoulder, arm, and wrist, considered to be related to her work involving the use of the Mouse: (1) calibration; (2) relaxing; (3) writing on the PC, using both hands; (4) using the Mouse with arm support; (5) using the Mouse while resting the arm on the table; (6) doing a variety of jobs. (From Rodahl et al., Project rapport, Norsk Hydro, Oslo, 1992d. With permission.)

suffer from such symptoms of neuromuscular complaints often are individuals who are especially sensitive to noxious stimuli of any kind, such as draught, temperature, noise, etc. At any rate, it should also be kept in mind that the symptoms in question may not necessarily be caused by pathological processes in the muscle tissue itself, but may possibly be due to the effect of muscle tension on other tissues, such as tendons, muscle attachments, etc., which are especially rich in pain sensors (for references see Åstrand and Rodahl, 1986).

BIOFEEDBACK

Not everyone is equally able to sense the state of tension of different muscle groups in their own body. In order to assist individuals in developing such

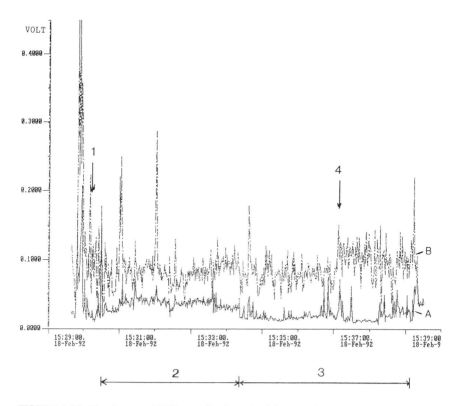

FIGURE 4-26. Simultaneous EMG recording from the right trapezius muscle (A) and from the right extensor digitorum muscle (B) in the same PC–Mouse operator as shown in Figure 4-25: (1) relaxing; (2) working with the Mouse, resting the hand on the Mouse; (3) working with the Mouse while resting the arm on the table; (4) telephone call. (From Rodahl et al., Project rapport, Norsk Hydro, Oslo, 1992d. With permission.)

muscular sensitivity, a variety of myoelectrical feedback instruments, such as the Myolog, have been developed. The function of myoelectric feedback instruments is to identify and amplify the myoelectrical voltage oscillations which appear over the muscle as it changes its state of tension and to present them as signals which the users can employ to improve control of their muscles. By setting a tension threshold combined with an audible signal, such an instrument can be used to let the subject know when the tension of the muscle being monitored, exceeds a certain limit, and to tell the person to relax.

The Myolog can be supplied with calibrated controls which determine the desired level of a signal required to obtain feedback. This can take the form of a beep feedback, proportional to the tension above threshold, or a display of a light signal when the muscle is completely relaxed.

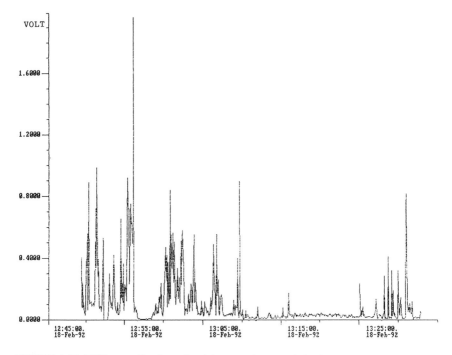

FIGURE 4-27. EMG recording from the right trapezius muscle in the same operator shown in Figures 4-25 and 4-26, when she was the center of attention in a television recording while doing her usual work with the PC–Mouse combination, and appearing to be visibly tense and nervous. Yet, she revealed no evidence of elevated muscle tension in her shoulder. (From Rodahl et al., Project rapport, Norsk Hydro, Oslo, 1992d. With permission.)

CHAPTER 5

The Monitoring of Heat Stress

The temperature of the environment is one of the factors affecting human performance (see Figure 2-7). At body temperatures substantially higher than the optimal levels (36.5 to 37.5°C), both physical and mental performance may deteriorate due to the complicated interplay of physiological and pathophysiological processes. Prolonged heat stress may lead to loss of body fluid (hypohydration), which in itself impairs performance, especially endurance. In addition, prolonged heat strain may impair mental and psychomotor functions, thereby affecting performance (for references, see Åstrand and Rodahl, 1986, Chapter 13). It is, therefore, of considerable practical importance to be able to assess the magnitude of the thermal stress in the working environment and the worker's physiological reaction to it, in order to ensure optimal conditions for health and productivity.

One of the most commonly used methods for the assessment of environmental heat exposure at different places of work is the Wet Bulb–Globe Temperature Index (WBGT), which is based on the measurement of the air temperature (dry bulb), radiant temperature (globe), and relative humidity (wet bulb). This can be metered and logged on the standard version of the Squirrel meter/logger. This, however, occupies three of the temperature channels of the logger. Furthermore, the WBGT index is rather complicated and the recording of it is time consuming. In the case of a predominantly dry, radiant heat environment, the much simpler and faster reacting Wet Globe Thermometer Index (WGT) or Botsball (Figure 5-1), devised by Botsford (1971), records similar values and the results are interchangeable with the aid of a simple formula (Ciriello and Snook, 1977; Beshir et al., 1982; Dernedde, 1992). Furthermore, the recording of the Botsball requires only one of the temperature channels of the Squirrel. The remaining channels can be used for the recording of physiological response to the heat stress by recording changes in body heat content assessed by rectal temperature (one channel), and mean skin temperature [indicated by the mid-thigh temperature (one channel)]. (For references, see Rodahl and Guthe, 1988.) Additional channels may be used for temperature parameters of special concern, such as radiant temperature (black ball temperature), eardrum temperature as an indication of central body

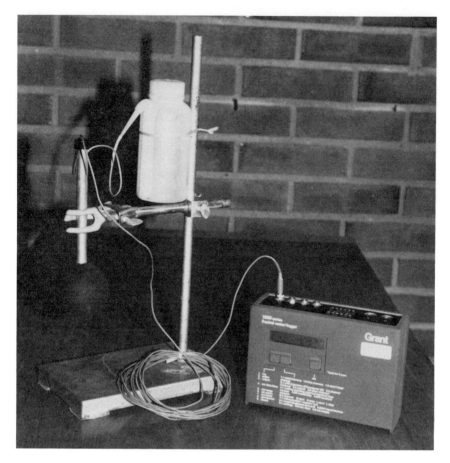

FIGURE 5-1. The Wet Globe Thermometer (also known as the Botsball thermometer), connected to a Squirrel meter/logger.

temperature, or the skin temperature (e.g., under the sole of the foot in a worker standing on a hot floor in front of a furnace).

THE BOTSBALL THERMOMETER

As already mentioned, the Botsball thermometer combines air temperature, humidity, wind, and thermal radiation into a single reading, expressing the thermal stress of the environment. The instrument is quite simple and easily adapted to industrial use. It consists of a small 60-mm hollow black globe, covered with a double layer of black cloth which is continuously moistened by water seeping from a reservoir tube attached to the globe. The temperature sensor attached to the Squirrel passes through a plastic tube along the center line of the water reservoir and into the globe, thus sensing its temperature.

THE MONITORING OF HEAT STRESS

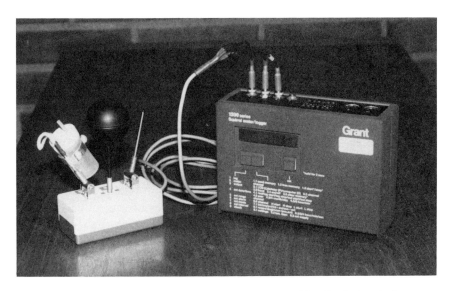

FIGURE 5-2. The Wet Bulb–Globe Thermometer, connected to a Squirrel meter/logger.

When placed in a hot area, the globe is heated by the surrounding air and by radiant heat from any hot surfaces in the surroundings. It is cooled by the evaporation of water from the globe surface, depending on air humidity and air movement. The wet globe reaches an equilibrium temperature when these heating and cooling effects come into balance, which usually takes about 5 min.

The Botsball temperature readings are in reasonably good agreement with the WBGT under conditions of dry, radiant heat. Since the former is easier to use and reacts faster, it may be considered preferable for practical field applications.

CONTINUOUS RECORDING OF ENVIRONMENTAL TEMPERATURE INDICES

The Squirrel logger, as already indicated, is ideally suited for the continuous recording of both the WBGT and WGT indices. The temperature probes for the Squirrel have been compared with a standard mercury thermometer and found to be quite accurate. The WBGT has three probes (Figure 5-2), occupying three channels in the Squirrel (1 = dry bulb temperature; 2 = globe temperature; 3 = wet bulb temperature). The WBGT index may be calculated by the following formula:

$$\text{WBGT} = (\text{Channel } 1 \times 0.1) + (\text{Channel } 2 \times 0.2) + \text{Channel } 3 \times 0.7)$$

The reading of the Botsball (WGT) can be converted to WBGT by the following formula:

$$\text{WBGT} = (\text{WGT}^2 \times 0.0212) + (\text{WGT} \times 0.192) + 9.5$$

The converted readings of the WGT have been compared with the WBGT readings, the two instruments being placed side by side in different places in and around the laboratory. The mean difference between the two sets of readings was 0.5°C, which is quite satisfactory. A similar comparison between the WBGT and WGT converted into WBGT was made in a ferroalloy plant in the north of Norway in the month of July. Here, under actual field conditions with changing radiant heat and considerable air movement, the mean difference was 1.5°C; the difference decreased as the temperature increased (range of difference 0.2 to 2.9°C). For all practical purposes, this degree of difference is quite acceptable in view of the fact that the environmental temperature in the plant may fluctuate by as much as 10°C from hour to hour in the same place.

In a more recent study, Dernedde (1992) showed that the WBGT can be predicted from the Botsball (BB in the following equation) temperature index and the ambient water vapor pressure (P_a), measured with a psychrometer, according to the following equation:

$$WBGT = 1.5157(BB) + 0.0112(BB \times P_a) - 0.7379(P_a) - 2.5591$$

Another advantage with the Botsball thermometer is that it can be mounted on the helmet and thus carried by the person, allowing the environmental temperature to be recorded at the actual location of the worker at any time (see Figure 5-3). The only drawback with this arrangement is that walking briskly causes the WGT to drop a couple of degrees due to air motion causing increased evaporation from the globe surface, hence the cooling of the probe. The ideal solution is probably to use the Botsball thermometer attached to the worker initially for the purpose of surveying the temperature distribution in the working area, and then to place the Botsball thermometer stationary at the location where the temperature is most extreme or most representative. The placing of the temperature probe inside the Botsball thermometer is not critical, providing the sensor is located within the globe of the Botsball.

ASSESSMENT OF HEAT STRAIN

It is clear that an assessment of the environmental heat stress can be made with the aid of the Squirrel logger using a WBGT assembly or a Botsball WGT instrument. However, the environment is one thing, but the human reaction to that environment is of far greater importance. For this reason, greater emphasis should be placed on the assessment of human response to the heat stress encountered by workers during the performance of their everyday work at their actual work places. Without such information any discussion of safe or upper limits of exposure would seem meaningless.

For the assessment of the physiological reaction to heat loads imposed by the environment, the body heat content (S) is a most meaningful index of body heat gain or body heat loss. It can be calculated by the following formula:

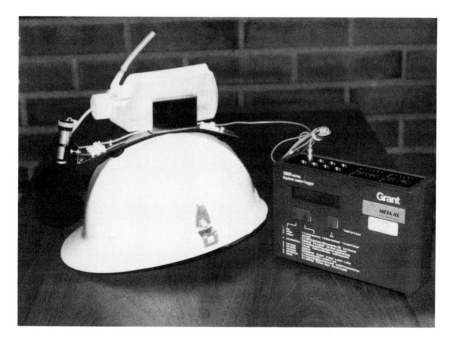

FIGURE 5-3. The Botsball thermometer (WGT), mounted on a helmet for ambulatory logging.

S (in kcal) = 0.83 × (body weight in kg) × (rectal temperature × 0.65) + mean skin temperature × 0.35).

Rectal Temperature

Rectal temperature mirrors the core temperature and, by itself, may reflect body heat gain or body heat loss. Under such conditions, rectal temperature may be used as an index of heat stress as long as sufficient consideration is given to the fact that it is a slow-reacting parameter. In most cases of industrial heat exposure, it takes some 45 min for the rectal temperature to reach a plateau. This emphasizes the importance of continuous recording of the rectal temperature by portable electronic loggers such as the Squirrel. This, however, requires cooperative subjects as well as patience and persuasion by the investigator. However, when correctly carried out, the continuous recording of the rectal temperature in workers in the course of their normal work and leisure is indeed a most revealing parameter in terms of thermal strain.

Ideally, the temperature sensor should be placed no less than 8 cm into the rectum in order to achieve stable readings representing deep body temperature. This may conveniently be done by supplying the rectal temperature sensor with a taped knob 8 cm from the end, to insert the sensor far enough into the rectum for the knob to be placed inside the rectal sphincter, and to tape the wire to the subject's lower back to prevent it from being pulled out of the rectum. For hygienic reasons, the sensor may be placed inside a thin disposable plastic envelope before insertion.

For practical reasons, it might, in some cases, be simpler to record the temperature inside the ear as an index of deep body temperature, replacing rectal temperature (T_{re}) in the body heat content equation. For a comprehensive discussion of the different deep body or core temperatures, the reader is referred to Chapter 13 in Åstrand and Rodahl (1986). For basic research purposes, it has been customary to measure the tympanic temperature as an expression of the deep body temperature, close to the location of the thermoregulatory center in the brain, by placing a thin thermocouple deep into the ear touching the lower part of the ear drum (Benzinger and Taylor, 1963). In the case of field measurements in working individuals at their industrial work places, this may not be any easier to do than using a rectal thermometer. The manipulation of the temperature sensor into the ear duct and the placement touching the ear drum requires some professional skill. In addition, it may be painful to the subject even when using a very thin and flexible thermistor wire. In any case, it is necessary to plug the external opening of the ear duct to prevent the ambient air interfering with the readings. (For references see Rodahl, K., 1989)

For the purpose of applied studies or less sophisticated research projects, Keatinge and Sloan (1975) have suggested a combination of an aural canal sensor and servo-controlled heating of the outer ear in order to attain oral temperature readings very close to ear drum temperatures.

For a rough estimation of deep body temperature, the much simpler oral (sublingual) temperature may be used, providing care is taken to keep the sensor or thermocouple in place under the tongue, and providing the subject is keeping his mouth completely closed. Oral temperatures recorded in this manner are, in most cases, about 0.5°C lower than the rectal temperature.

Skin Temperature

Skin temperature can be measured relatively easily with the aid of commercially available skin temperature sensors connected to the Squirrel. It may be used as an index of local heat exposure, but cannot by itself be used as an indication of general heat stress or thermal balance. Proper shielding of the temperature sensor under conditions of intense radiating heat is essential in order to obtain real values of skin temperature. Otherwise, it should be noted that the skin temperature recorded on the inside of the thigh closely resembles mean skin temperature under most normal circumstances (Ramanathan, 1964). This value may, therefore, be used in the body heat content equation previously referenced.

From this brief description, it is evident that the miniature electronic Squirrel logger, which can easily be carried on the individual, is ideally suited for the continuous logging of both the environmental heat stress (in terms of the WGT) and the resultant heat strain (in terms of changes in body heat content), reflected by changes in core and shell temperatures. Furthermore, all these parameters may be recorded simultaneously by the same Squirrel logger. With the new series of multichannel Squirrel loggers a number of other pertinent parameters in addition to temperature, such as heart rate (as an indication of work stress), muscular tension, concentration of dust and different gases in the ambient air, etc., can be continuously and simultaneously recorded on the same logger.

We shall now present a series of examples showing the results of the application of the Squirrel logger–temperature sensor combination in a few industrial work places:

HEAT STRESS AND HEAT STRAIN IN A TYPICAL NORWEGIAN ALUMINUM PLANT

At the request of the manager of the Health and Safety section of a major aluminum production plant in the northern part of Norway, a systematic assessment was made of the heat stress to which some of the pot room operators were exposed and the effect of the exposure on the operators (Nes et al., 1990). The survey which took place in 1989 and 1990 was based on the combination of the 12-bit Squirrel logger and the standard temperature probes for skin and rectal temperature in addition to the Botsball and the black ball thermometers. The study was performed in close collaboration with the engineers of the plant laboratory. The actual logging operation was gradually taken over by one of the engineers and eventually developed into an advanced logging capability within the plant, involving a variety of environmental parameters besides temperature.

The study was in accordance with the recently established Norwegian regulations which requires the industry itself to monitor its own environmental health hazards. The purpose of this study was to visualize the actual heat stress involved in the different operations as a basis for a meaningful discussion of possible improvements with the aim of enhancing the health and well-being of the worker which, in turn, might also enhance productivity.

A total of eight operators, of whom two were women, took part in the study. Three of them were engaged in gas-manifold changing, three in burner cleaning, one in Jack raising, and one working in the foundry. They were studied during a total of 38 work shifts, both during the winter and the summer. Each subject was studied repeatedly during subsequent days in order to determine the reproducibility of the results.

The ambient heat stress was recorded by the Botsball thermometer mounted on top of the worker's helmet. This made it possible to measure the ambient heat stress at the place where the operator was actually working. In addition, a small black-ball thermometer, measuring the radiant heat, was placed next to the Botsball thermometer on the helmet.

The physiological reaction to the heat stress in terms of body temperature was recorded—i.e., skin temperature on the middle of the thigh as an indication of mean skin temperature and the rectal temperature as an expression of core temperature. This was done with the aid of temperature sensors (supplied by Grant Instruments Ltd.) plugged into the temperature inputs of the 12-bit Squirrel logger. The logger was shielded against the magnetic field in the pot room by being kept in a fitted steel box and carried in a belt on the subject's back. The stored data was transferred to a personal computer and the results displayed immediately after the observation period to those involved in the study. One advantage with this immediate display of the results was that the subject could see the effects of his different activities in terms of specific heat stress and the direct relationship between cause and effect. The fluid loss in terms of sweating was determined by weighing the subject on an accurate scale, before and after the work shift. The weight of food and fluid intake as well as the stool and urine output were taken into consideration.

The results of this project showed convincingly that heat stress was a major problem at this particular aluminum plant. In this respect, the results confirmed the findings of a previous study of an aluminum plant in the south of Norway which was performed with less sophisticated equipment a number of years before this study (see Chapter 1, Figure 1-9). The results showed considerable differences in heat stress at the different plant operations, as well as variations in the heat stress from one day to the next, and at different seasons of the year (Nes et al., 1990). From the observations made, it was evident that burner cleaning represented by far the most severe heat stress. It had Botsball temperatures as high as 35°C and radiant temperatures exceeding 60°C. The rectal temperature exceeded 38°C in all the subjects studied, in some cases with peaks over 39°C (Figure 5-4). The skin temperature oscillated with the radiant temperature exceeding 40°C and in some cases, reached peaks close to the pain threshold (Figure 5-5).

Next in terms of heat stress was gas manifold changing, where in one case the skin temperature under the foot reached about 42°C (Figure 5-6). Then came Jack raising on the prebaked pots. In one case, this involved continuous exposure for more than 2 hours, causing a gradual rise in the rectal temperature from 37°C to more than 38°C in spite of a moderate Botsball temperature from 15 to about 25°C.

The weighing of fluid loss and hypohydration due to sweating reflected, on the whole, the magnitude of the heat stress as evidenced by the Botsball and body temperature. The mean values showed that those who were most exposed to heat stress also had the greatest fluid loss by sweating. The fluid loss in the same subjects (examined both during summer and winter) showed a significantly lower fluid loss during the winter (Nes et al., 1990), but even in the winter the fluid loss exceeded the levels which, as a rule, may give symptoms of hypohydration with fatigue, reduced stamina, and reduced alertness (for references, see Åstrand and Rodahl, 1986).

In some of the subjects, a negative fluid balance, i.e., a net deficiency in fluid intake of 3.5 L during a work shift, was recorded. This represents a significant degree of hypohydration and emphasizes the need of the operators to drink water regularly during the work shift, regardless of whether or not they feel thirsty. This is due to the feeling of thirst being a slow-reacting indicator of the body's state of fluid balance. In addition, sweat loss may be significantly reduced by reducing each period of heat exposure, e.g., 20 min, interspersed by frequent 10-min cooling-off periods in order to prevent an excessive rise of the internal body temperature. The heat strain could be significantly reduced by using heat resistant

THE MONITORING OF HEAT STRESS

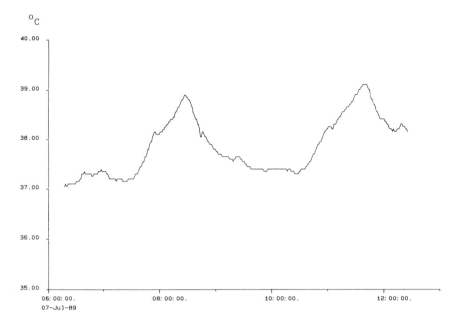

FIGURE 5-4. Rectal temperature in a burner cleaner in an aluminum plant, with peaks exceeding 39°C. (From Nes, H., R. Karstensen, and K. Rodahl, Varmestressundersøkelse ved Elkem Aluminum, Mosjøen, Technical Report, 1990. With permission.)

and heat reflective clothing, including reflective aprons. The need for effective protective garments is particularly stressed by the finding of the very high skin temperatures both on the legs and under the foot soles. Some of our figures in these areas approached the level of activating pain sensors in the skin. These findings merely confirmed the experience of the operators, and may easily be remedied by using reflective shields in front of their legs and heat resistant insulating insoles and perhaps even covering the boots with heat-reflective material, as well as using heat resistant gloves or mittens.

HEAT STRESS AND HEAT STRAIN IN THE MODERN GLASS INDUSTRY

In a major glass bottle factory in the south of Norway, both management and labor were concerned about the high incidence of minor accidents involving hands and fingers among the heat-exposed operators. They wondered about the possibility of ambient heat stress affecting the operators' mental alertness, which, in turn, might be the cause of minor accidents, especially of the hands and fingers. This was the reason for a series of extensive studies in this plant, which involved the assessment of the ambient heat stress and its effect on workers in a variety of job locations. This project was extended to investigate possible protective measures in order to reduce the heat strain of the operator, including adequate water intake and the use of protective garments and other protective devices.

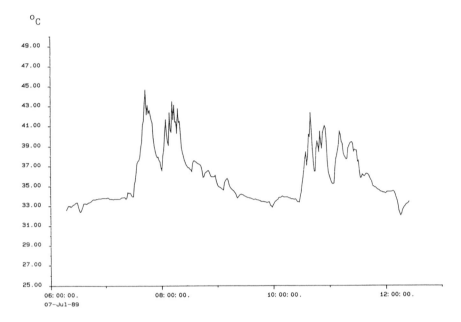

FIGURE 5-5. Skin temperature on the inside of the thigh in a burner cleaner in an aluminum plant, approaching the pain threshold. (From Nes, H., R. Karstensen, and K. Rodahl, Varmestressundersøkelse ved Elkem Aluminum, Mosjøen, Technical Report, 1990. With permission.)

In an initial survey in September 1989, three subjects were followed in the course of their normal daily work: one during a routine inspection of two different production lines, including the necessary mechanical maintenance work of the actual machinery; one subject operating a bottle production line, including the changing from the production of one bottle size to another; the third subject was an electrician involved in different maintenance operations on top of and between the ovens (Rodahl et al., 1989).

It appeared that the heat stress varied greatly from one production line to the next, depending, to a large extent, on the size of the bottles being produced: the larger the bottles, the greater the emission of radiant heat. The Botsball thermometer mounted on top of the helmet of the first subject, the smelter, recorded values as high as 40°C at the first inspected oven. There was a gradual increase in rectal temperature from about 37°C to more than 38°C. The second subject, operating one of the production lines, was exposed to a more constant but less extreme heat stress, with Botsball temperatures between 25 and 30°C, causing a gradual increase in the rectal temperature to 38°C. When working on the top of and between the ovens, the electrician was exposed to greatly varying heat stress, with Botsball temperatures up to 30°C, causing skin temperatures around 36°C, and rectal temperatures close to 38°C.

Thus, the study showed that such a glass factory may represent considerable ambient heat stress. It also showed that the simplicity of the logging equipment

THE MONITORING OF HEAT STRESS

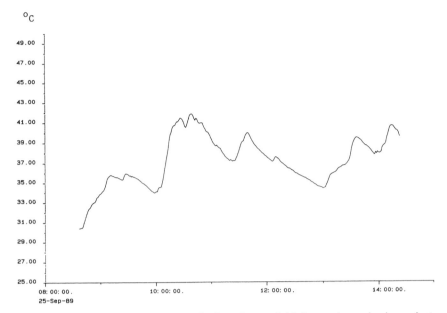

FIGURE 5-6. Skin temperature under the foot of a manifold changer in an aluminum plant. (From Nes, H., R. Karstensen, and K. Rodahl, Varmestressundersøkelse ved Elkem Aluminum, Mosjøen, Technical Report, 1990. With permission.)

used made it possible for the technical staff to perform most of the measurements and loggings by themselves.

In a subsequent, more extensive study at the same plant (Nitter et al., 1990), seven key operators covering different functions of the glass bottle production were studied during their entire work shifts. Most of the actual logging and data collections were done by the members of the technical staff of the plant. For the logging of the ambient heat stress, a Botsball and a miniature black-ball radiation thermometer were mounted on the helmet worn by the subject. The subject's physiological heat reaction to the heat stress, in terms of body temperature, was measured with the aid of a rectal temperature sensor and a skin temperature sensor placed on the inside of the thigh. All sensors were connected with the Squirrel ambulatory logger carried in a belt by the subject. The fluid balance was recorded in the usual manner by the changes in body weight and the weighing of fluid intake and output.

An example of the typical heat stress in front of the melting platform at the glass bottle production line is given in Figure 5-7. The heat exposure was fluctuating from 10 to 25°C; in general, below the recommended level of 25°C. The skin temperature was relatively moderate, below 35°C, and the rectal temperature showed a gradual rise from about 37°C to about 37.8°C, with a clear dip during the breakfast break.

The fluid balance clearly reflected the heat stress to which the individual operators were exposed. Five of the seven subjects had a net fluid loss of about

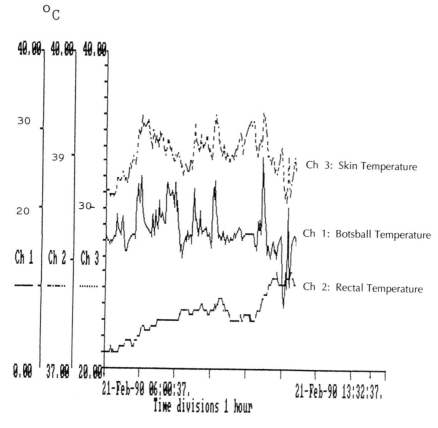

FIGURE 5-7. An example of the typical heat stress, as evidenced by Botsball, skin, and rectal temperature, in an operator working in front of the melting platform at a glass bottle production line. Channel 1: Botsball temperature; Channel 2: rectal temperature; Channel 3: skin temperature. (Nitter et al., februar–mars, 1990, KIL-amil-dok-5. March 29, 1990. With permission.)

1 L or more in the course of the work shift, corresponding to more than 1% of the body weight, or close to the order of magnitude when the effects of hypohydration might be noticeable.

This project clearly demonstrated that it is possible for the industry itself, with the aid of their own staff, to properly log the heat stress to which their operators are exposed and its physiological effects.

It also revealed that, at a glass manufacturing plant of this kind, the heat stress may be sufficiently high to cause physiological effects in terms of increased body temperature and hypohydration of a degree sufficient to cause fatigue, reduced endurance, reduced alertness, etc.

In view of the fact that the major cause of the heat stress in this type of industry is radiant heat, a great deal can be achieved simply by using heat-reflecting protection, for instance, in the form of a reflective apron. The observed hypohydration can be eliminated by encouraging the operators to drink water in adequate quantities at regular intervals during the working hours. The effect of a reflective apron has already been demonstrated in a study in a Norwegian ferroalloy plant where a loosely fitted aluminum-covered reflecting apron was worn by the subject when tapping the pots (Figure 5-8).

A year later, another heat stress study was carried out in a major Norwegian crystal glass producing plant. The study consisted of an ambulatory recording of the heat stress and the heat strain of four operators involved in the production of crystal table glasses (Guthe et al., 1990). The purpose of the project was to attain an objective visualization of the heat stress to which these operators were exposed in the course of a work day in the glass-producing part of the plant.

The methods and the procedures used were the same as those used at the previously mentioned glass factory, using the Botsball thermometer mounted on the helmet for the recording of the ambient heat stress, and recording the skin and rectal temperature as an indication of heat strain. All the recorded parameters were logged by a 1202-series Squirrel logger. As usual, the operators that were studied used ordinary work clothes without any heat protection.

The results obtained in the study of the first subject (Figure 5-9) may be taken as fairly representative for all four subjects studied and involved some of the ordinary everyday operations. The subject started by carrying trays of finished hot crystal candlestick holders from the production line to a cooling-off cabinet. Following a brief coffee break, he started an operation known as "heating" which normally requires 15 min but which, in this case, was prolonged an additional 20 min in order to demonstrate the time-related accumulation of body heat and to what extent it might level-off in the course of time. The operation involved the operator dipping a 1.5-m-long iron pipe through a small opening into an oven containing melted glass, in order to pick up a small glass blob at the end of the pipe. In so doing, the operator was exposed to a considerable amount of radiant heat. This was repeated about once every minute or so. With the glowing glass blob at the end, the pipe was continuously rotated while being carried a few yards to a table where the glass blob was dropped into a candlestick holder form. As is evident from Figure 5-9, this procedure represented considerable heat stress. In a matter of minutes, the Botsball temperature rose to above 30°C. This caused the skin temperature of the thigh to rise to between 39 and 40°C, and was mainly due to the fact that the subject was only wearing ordinary trousers which did not reflect the radiant heat. The rectal temperature rose to above 38°C after about 20 to 25 min. It should be mentioned that this indicates that with an exposure period of 15 min, which was normally the case, the subject's rectal temperature, in all probability, would have remained below 38°C. This indicates that the standard routine to which the worker normally adhered probably would be acceptable from a physiological standpoint.

This survey did show that some of the operators at this particular glass factory may be regularly exposed to considerable heat stress which may be sufficient to cause considerable elevation of body temperature and, as a result,

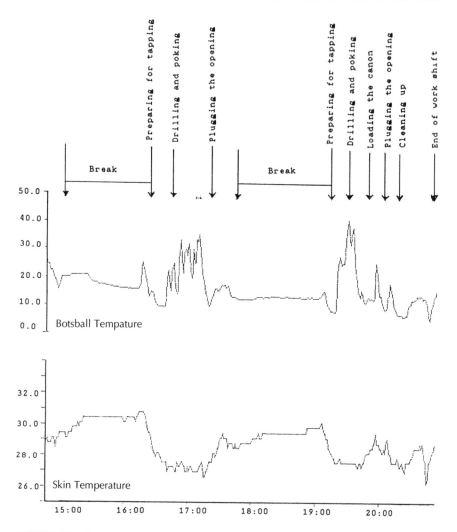

FIGURE 5-8. The effect of a loosely fitted, aluminum-covered reflective apron on the skin temperature of the thigh in an operator in a ferroalloy plant, showing a drop in the skin temperature (bottom) in spite of a rise in the Botsball temperature (top) when tapping the pots. The reason for the drop in the skin temperature may be the cooling effect of the evaporation of the sweat.

cause persistent sweating and some degree of hypohydration with its negative physiological effects. On the other hand, the observations made do show that varying the work procedures, i.e., rotating from one operation to another at regular intervals, is a reasonable solution to the heat stress problem. Nonetheless, the radiant heat exposure may be sufficiently severe to warrant a systematic use of protective clothing and other measures: especially reflective aprons or reflective covers on the work clothing, including gloves or mittens, and protective shielding of the face, as well as regular fluid intake in order to avoid hypohydration.

THE MONITORING OF HEAT STRESS

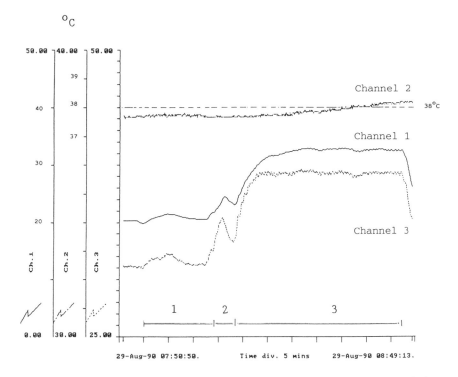

FIGURE 5-9. An example of the heat stress exposure of a worker in a crystal glass production plant: (1) carrying finished, hot crystal candlestick holders from the production line to the cooling-off cabinet; (2) coffee break; (3) transferring small glass blobs at the end of a pipe from the oven to the candlestick holder form. Channel 1: Botsball temperature; Channel 2: rectal temperature; Channel 3: skin temperature.

EFFECT OF ADEQUATE FLUID INTAKE

As a practical consequence of the results obtained in some of these studies, the effect of added fluid intake on body temperature and heart rate was investigated in one operator at the glass bottle–producing plant. The operator went 1 day without and 1 day with fluid intake (250 mL water every 30 min) under conditions comparable to the previously mentioned study (Rodahl et al., 1991c). (For references concerning the effect of a negative fluid balance, see Åstrand and Rodahl, 1986, Chapter 13.) The subject was exposed to a 10-min standard work load of 600 kpm/min on a cycle ergometer at normal room temperature, before and after the regular work shift, 1 day with fluid and 1 day without fluid intake.

On the day without fluid intake, he had a net fluid loss due to sweating of 1.9 L. On the day with fluid, his fluid intake exceeded his fluid loss in terms of sweating. On the day without fluid intake, his mean heart rate during the ergometer exercise was 5 beats/min higher after than before the work shift. On the day with the fluid intake, it was 13 beats/min lower after than before the work shift.

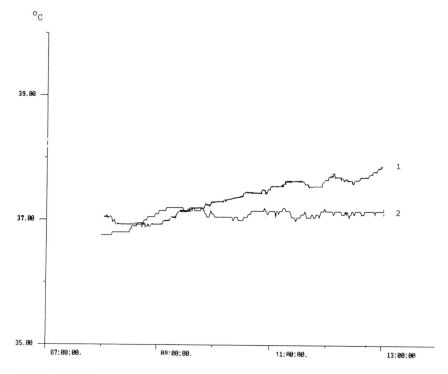

FIGURE 5-10. The effect of adequate fluid intake on the rectal temperature of an operator in a glass bottle production plant: (1) without fluid; (2) with fluid.

The mean rectal temperature during the 10 min work test was 0.6°C higher after than before the work shift in the case where no fluid was taken vs. only 0.3°C higher when adequate fluid was taken. The minute to minute changes in the rectal temperature with and without fluid are shown in Figure 5-10.

These findings are in agreement with previous observations made at a Norwegian cement factory, in which case the mean heart rate at a similar standard cycle ergometer work load at the end of the work shift was 10 beats/min lower in the subjects consuming 2 L of water during the work shift, than when they consumed no fluid during the work shift (unpublished results, 1975).

HEAT-PROTECTIVE CLOTHING

Another practical consequence of the general industrial heat stress studies which were made with the Squirrel logger–sensor combinations, was the awareness of the need for more detailed investigations of the possibilities of providing protective measures. This includes protective garments which shield the operators against the untoward effect of the ambient heat to which they are exposed.

In the above-mentioned glass factory, an aluminum-reflecting cover glued to ordinary woolen material was taped to the right leg of the operator's work pants

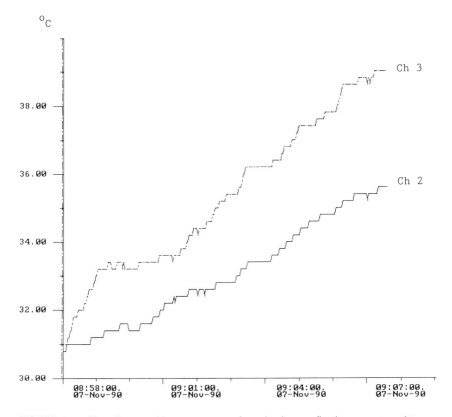

FIGURE 5-11. The effect on skin temperature of an aluminum reflective cover taped to one of the legs of an operator in a glass factory: (Ch 2) with reflective cover; (Ch 3) without reflective cover.

with the reflecting side facing outward. This was done so that the reflecting material covered the right thigh. The left leg was unshielded. Skin temperature sensors were taped on both thighs, and connected to an 8-bit Eltek Squirrel logger. The subject stood in front of one of the production lines where the radiant temperature exceeded 50°C. As expected, the skin temperature of the unshielded leg increased significantly more than that of the shielded leg (Figure 5-11) (Rodahl et al., 1989).

In a subsequent experiment, the shielding effect of a reflective apron was compared to that of a nonreflective, protective apron based on the skin temperature of the thigh (Rodahl et al., 1990). In addition to a skin sensor placed in the middle of the right thigh, a black-ball radiation thermometer was placed on the inside of the apron. The subject wore ordinary work pants and stood on a wooden box in front of the production line so that the middle of his body, including his thighs, was exposed to the radiant heat source.

As is evident from Figure 5-12, the skin temperature on the thigh behind the reflective aluminum apron increased significantly less than did the skin temperature on the thigh behind the nonreflective skin apron.

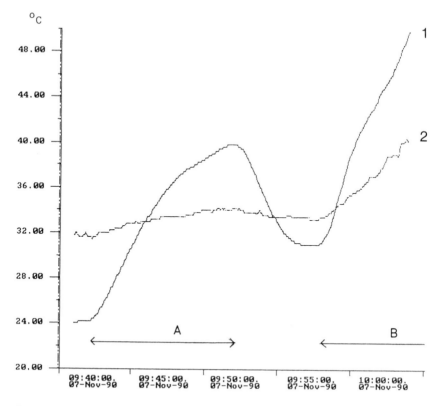

FIGURE 5-12. The shielding effect of a reflective apron (A), compared to that of a nonreflective apron (B), on the skin temperature of the thigh, (1) black-ball radiant-heat sensor placed behind the apron, (2) skin temperature of the thigh.

The effect on the skin temperature of thin aluminum-covered insoles under the foot was tested in a subject standing on top of one of the glass-producing ovens, where the temperature was extremely high. The insole, with the aluminum facing down, was placed inside the right shoe. In the left shoe, there was no insole. Skin sensors connected to the Eltek logger were taped to the skin under both the left and the right foot sole. The subject wore work clothing and a pair of ordinary socks and work shoes. He stood on the metal floor on top of one of the glass-producing ovens for a period of 10 min.

From Figure 5-13 it is observed that the skin temperature under the left foot sole increased much more than that under the right foot sole. However, even with the aluminum reflecting insole, the temperature increased considerably, probably due to the conductive heating of the soles of the shoes which were in direct contact with the hot metal floor. It would therefore seem advisable to use a thicker insulating insole covered with a thin aluminum shield inside shoes covered with a reflective material to reduce the heating of the shoes.

These observations were extended by a series of measurements at the same glass factory, using the same subject (Rodahl et al., 1992c). The ambient temperature

THE MONITORING OF HEAT STRESS

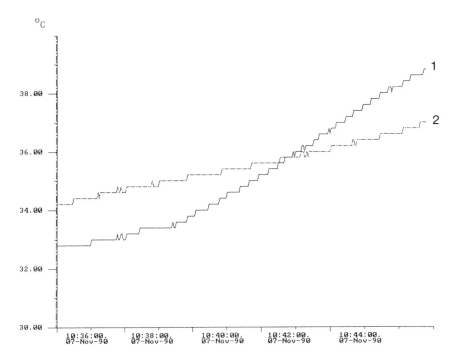

FIGURE 5-13. The effect of thin aluminum-covered insoles on the skin temperature under the foot in a subject standing on top of a glass-producing oven: (1) without insole; (2) with insole.

was measured with a dry-air temperature sensor, and the radiant heat by a black ball thermometer connected to the 12-bit Squirrel logger. The temperature sensors were mounted on the subject's helmet. The skin temperature, as well as the temperature of the clothing and the head protection, were measured with the aid of thermistors plugged into the logger.

As can be seen from Figure 5-14, the air temperature and the radiant temperature to which the subject was exposed, when standing on a wooden box in front of the glass bottle production line, were about 50°C (mean 49.0°C) and 70°C (mean 63.6°C), respectively.

When wearing a specially made reflective apron made of vacuum-treated trioxypan fibers, the skin temperature on the abdomen was below 36°C (Figure 5-15), but rose to about 41°C without the apron (Figure 5-16). Using an ordinary, disposable, aluminum-covered reflective apron, the abdominal skin temperature was only about 34°C (Figure 5-17), indicating that the ordinary, disposable, aluminum-covered reflective apron was as good or even better than the more elaborate amide fiber apron. It should also be pointed out that the temperature on the outer surface of the shiny aluminum-covered reflective apron was considerably lower (around 50°C, Figure 5-17), than in the case of the more elaborate aluminum fiber apron (approaching 70°C, Figure 5-15). This may be because the shiny surface of the aluminum apron reflected the radiant heat from the red hot

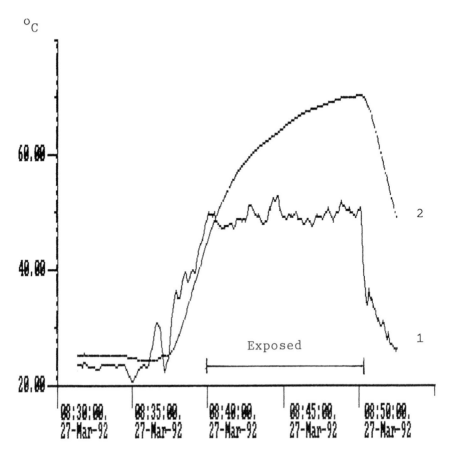

FIGURE 5-14. The ambient heat stress in terms of dry air temperature (1), and radiant temperature (2), recorded by sensors mounted on the subjects helmet, 1.5 m away from the row of glowing glass bottles on the production line.

glass bottles to a greater extent than did the fibers. Figure 5-18 shows the effect of shielding the arms against the radiant heat by aluminum fibers in the same subject standing in front of the same production line.

With a reflective visor in front of the face, the effect on the forehead skin temperature was measured in the same subject exposed to the same heat stress. The forehead skin temperature rose considerably in spite of the reflective visor. This is probably because the air temperature behind the visor was high enough to raise the temperature of the skin, even without the radiant heat exposure.

PROTECTION OF THE HEAD

As has been repeatedly pointed out, there are a number of observations indicating that the head plays a key role in the temperature regulation of the body as

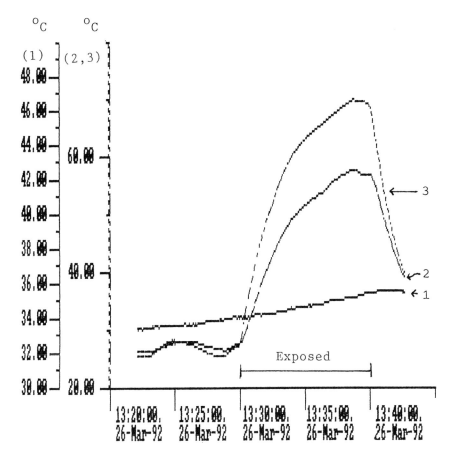

FIGURE 5-15. The effect of a reflective apron made of vacuum-treated trioxypan fibers, on the skin temperature of the abdomen in a subject exposed to the heat stress of a glass bottle production line. The rise in skin temperature is quite moderate: (1) skin temperature on the abdomen; (2) temperature on the inside of the apron; (3) temperature on the outside of the apron.

a whole. This was discovered during the space research activity, when the effect of the cooling of the head on the temperature regulation of the body was revealed, as well as it's effect upon the subject's subjective conception of the heat stress (for references, see Åstrand and Rodahl, 1986; Rodahl, K., 1989).

One has also become aware of the fact that the head has a vascular system which is especially well suited for the cooling of the brain (Cabanac, 1986). Our own recent studies have shown that the use of bicycle helmets causes heating of the head. They have also shown that when the head is heated above a certain limit, the subject's psychomotor precision is reduced. This may in turn lead to a greater risk for accidents due to reduced alertness, etc. (Bjørklund et al., 1991).

At several Norwegian industrial work places, excessive heat stress of the head was recorded in the course of our recent heat stress survey. This also applies to our glass factories, where the heat stress may be particularly high on top of the glass-producing

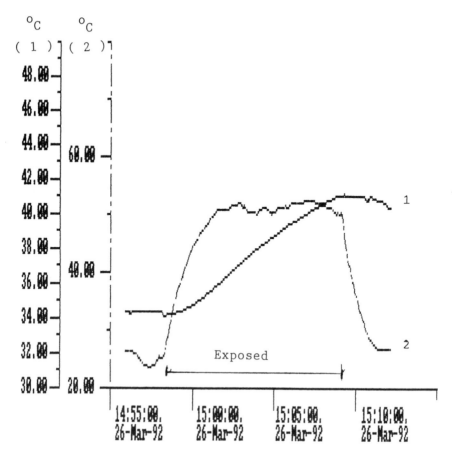

FIGURE 5-16. The marked rise in skin temperature of the abdomen in a subject exposed to the same heat stress as in Figure 5-15, without any heat-protective clothing: (1) skin temperature on the abdomen; (2) air temperature.

ovens and where some of the maintenance operators at times may be working for varying periods of time involved in such operations as changing electrodes, etc. (Rodahl and Guthe, 1991). Here the radiant temperature may be in the order of 70 to 80°C, and the air temperature between 80 and 90°C. Under conditions such as these, the skin temperature on the middle of the thigh, even though it is covered by the worker's pants, may reach temperatures above 40°C, and the skin temperature of the head may exceed 44°C, even at the end of a 10-min exposure.

In view of the significant role played by the head in the physiological reaction to heat, protective measures to prevent the overheating of the head would be a matter of high priority. One such protective measure would be the use of one of the existing ventilated helmets available on the market, such as the Racal Airstream helmet produced by Racal Safety Ltd., Wembley, Middlesex, England. The fan of such a helmet is operated by a rechargeable pocket battery. The air is blown down over the face. The helmet is equipped with an air filter and a plastic

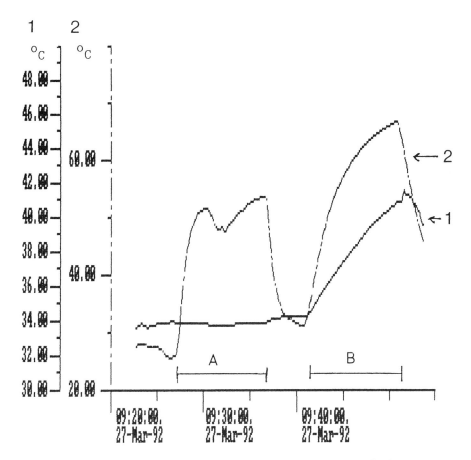

FIGURE 5-17. The effect of an ordinary, disposable, aluminum-covered, reflective apron on the abdominal skin temperature: (A) with apron; (B) without apron. The subject is exposed to the same heat stress as in Figure 5-15 and 5-16. (1) abdominal skin temperature; (2) temperature on the outside of the apron (A), or on the outside of the jacket (B).

visor as an eye protection and an additional metal visor on the outside as a protection against scratching, sparks, or heated particles in the air.

This helmet was used by an experienced operator seated on top of one of the glass-producing ovens (where the air and the radiant heat were between 70 and 80°C) and standing in front of one of the glass bottle production lines (where the air and radiant temperatures were between 40 and 50°C). The temperature parameters were recorded by an ambulatory Squirrel logger. The heat stress sensors were fixed to the helmet which was carried by the subject, either on his head or by hand.

The results showed that the skin temperature on the subject's forehead approached 41°C while sitting on the top of the oven with his helmet in place and the fan switched on. Practically speaking, this is actually the same temperature of his forehead as when using an ordinary helmet. The reason for the apparent lack

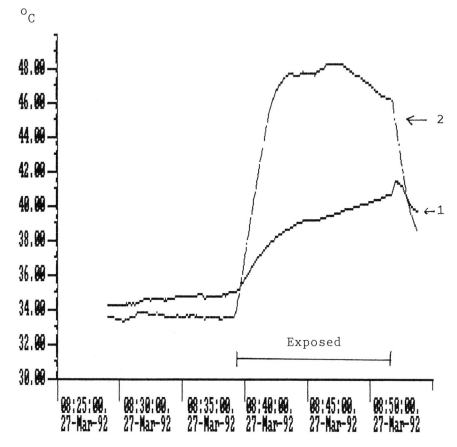

FIGURE 5-18. The effect of shielding the arms against the same radiant heat stress as in Figures 5-15, 5-16, and 5-17, by aluminum–reflective fiber material. (1) skin temperature of the protected arm, (2) skin temperature of the unprotected arm.

of effect of the fan of the helmet may be the fact that the air which is blown across the face is very hot.

Figure 5-19 shows the temperature of the head of the subject, standing in front of the glass bottle production line, 1.5 m away from the heat source, bareheaded (A), wearing the ventilating helmet with the fan operating (B), and finally, wearing an ordinary helmet (C). It is quite evident that the ventilating helmet, as long as the fan is active, provides the best cooling effect of the head temperature.

The Effect of Protective Helmets

It is generally agreed that the use of helmets for the protection of the head is desirable, both in cyclists and in exposed industrial workers. The use of such helmets, however, may cause the head to become too hot. This may affect the

THE MONITORING OF HEAT STRESS

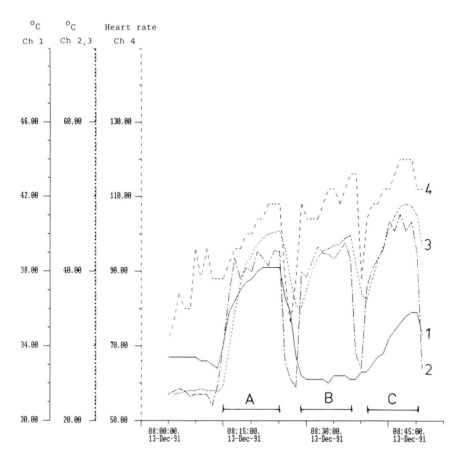

FIGURE 5-19. The cooling effect of a ventilating helmet, on the forehead skin temperature, in a subject standing in front of a glass bottle production line: (A) bareheaded; (B) wearing the ventilating helmet; (C) wearing an ordinary helmet; (1) Ch 1, forehead skin temperature; (2) Ch 2, air temperature; (3) Ch 3, radiant temperature; (4) Ch 4, heart rate.

thermal regulation of the entire body, as has been repeatedly pointed out in this chapter (for references see Åstrand and Rodahl, 1986; Rodahl, K., 1989).

It has been shown that the wearing of a protective face mask imposes a significant additional heat stress, causing increased sweat rate, elevated rectal and skin temperature, and increased heart rate (Martin and Callaway, 1974). On the other hand, cooling of the skin of the head brings about reduced sweating over the entire body, evidently mediated through a change in hypothalamic temperature (Baker, 1982). In subjects exposed to heat, Morales and Konz (1968) showed that the use of a water-cooled hood caused not only a drop in the head temperature, but also in the core temperature. It caused reduced sweating and enabled the subjects to endure the heat stress for a longer period of time. Konz and Gupta (1969) showed that localized cooling of the head during heat exposure in the form of a cooled hood caused a reduced sweat rate and less increase in the heart rate, as well as less decline in mental performance. These effects have essentially been confirmed by Shvartz (1970, 1976), Nunneley et al. (1970), Greenleaf et al. (1980), and Brown and Williams (1982).

During physical work, the heat production increases proportionally to the energy output causing an increase in body temperature, most marked in the muscles. The temperature inside the head, i.e., the brain, does not normally rise as high as in the rest of the body. Evidently, this is due to a special arrangement of the vascular system in the head which provides a selective cooling of the brain (Cabanac, 1986). When the hyperthermia is caused by prolonged exercise, such as running or bicycling without a helmet, the movement of the air around the head is increased. This facilitates the cooling of the head by increasing the evaporation of the sweat. Cabanac (1986) has pointed out that bald individuals may have the advantage of additional heat loss due to sweating from the hairless scalp because sweating from the bald scalp can be as great or greater than that of the forehead. Enclosing part of the head in a helmet may have the opposite effect causing an excessive heating of the head because the helmet may hamper the evaporation of the sweat and consequently reduce heat dissipation. This may be particularly undesirable since brain function appears to be especially vulnerable to heat (Baker, 1982). Pepler (1963) and Wyon et al. (1979) have reported deterioration in mental performance in subjects exposed to heat stress. This may be of particular importance to cyclists, who depend on mental alertness, balance, and integrated mental and physical performance. Similarly, it may be of importance to heat-exposed industrial workers, who depend on mental alertness to avoid accidents. While there is a vast number of publications dealing with the use of bicycle helmets in relation to safety, accidents, injuries, etc. (e.g., Thompson et al., 1989), a review of the current literature on this subject reveals that there is very little data available showing the effect of the use of bicycle helmets on head temperature and its consequence in terms of psychomotor function.

On this basis, we investigated the effect of the use of helmets on the body temperature under controlled laboratory conditions and under actual field conditions. We particularly investigated the temperature of the head and to what extent changes in head temperature affect psychomotor performance (Rodahl et al., 1992e). This project was based on the use of the Squirrel logger, carried on the subject, combined with available sensors for ambient and body temperature and relative humidity, recording six parameters simultaneously (Figure 5-20). The psychomotor performance tests consisted of a simple choice reaction-time test and a complex choice reaction-time test (Bjørklund et al., 1991). Six subjects were included in the laboratory tests and four of them in the bicycle field test.

From a collection of helmets which complied with the safety requirements set by the Norwegian Product Safety authorities, a typical closed helmet was chosen and used in all basic experiments. In a few separate experiments, an "open" helmet was also used. An electronic relative-humidity sensor was placed above the subject's head, inside the helmet, and connected to the Squirrel ambulatory logger for continuous recording of the relative humidity inside the helmet.

THE MONITORING OF HEAT STRESS

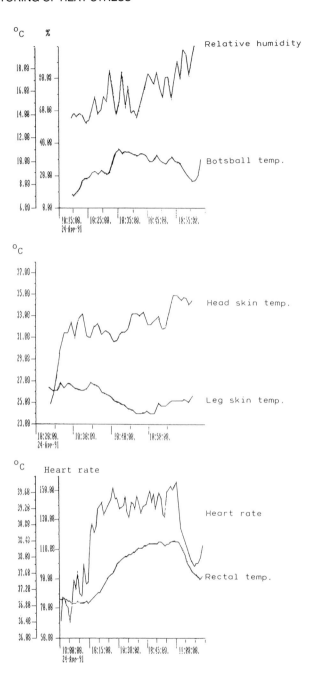

FIGURE 5-20. The Botsball temperature, relative humidity under the helmet, skin temperature of the forehead and of the leg, rectal temperature, and heart rate, in a subject during a 60-min outdoor bicycle ride, using a closed bicycle helmet (Rodahl et al., The Fifth International Conference on Environmental Ergonomics, Maastricht, Nov. 2-6, 1992e. With permission.)

The laboratory experiments were carried out in the climatic chamber at the National Institute of Occupational Health in Oslo. The room temperature was kept between 20 and 22°C. The sensors recorded rectal temperature, skin temperature at the middle of the thigh (an approximate indication of mean body skin temperature), the skin temperature of the head, the heart rate, the relative humidity inside the helmet, and the Botsball temperature outside the helmet. The sensors were attached to the subject. In a series of initial experiments, it was established that the head skin electrode should be placed at the upper left or right corner of the forehead in order to attain the highest head skin temperature of the part of the head covered by the helmet, without having to remove part of the subject's hair. Following the attachment of the sensors, the psychomotor tests were performed twice.

The subject then exercised on the Monark Cycle ergometer, starting with a warm-up period of 5 min at 50 watts. The rate was kept at 60 revolutions per minute. This was immediately followed by a 60-min cycle ergometer exercise at a work load varying from 75 to 150 watts, depending on the physical fitness of the subject. All subjects were accustomed to bicycling. Following the ergometer exercise, the psychomotor tests were repeated.

A series of field tests were performed in four of the subjects, consisting of 1-hour bicycle rides with and without a helmet. The subjects reported to a field laboratory established for this purpose in a building high on a hill outside Oslo. Here, following attachment of the sensors, etc., the subjects completed the psychomotor tests and bicycled with the closed helmet down the hill and into a valley to a fixed turning point, and back again. The work rate was kept at a rather high level, and the subjects were followed by a cyclist or by a car for safety. The ambient temperature varied between 6 and 16°C (Botsball temperature 6.5 to 16.7°C). Immediately after returning from the bicycle ride, the psychomotor tests were repeated, and the subject was allowed to rest for about 10 min. Following this, the subject repeated the same bicycle ride, but this time without a helmet, followed by a final psychomotor test.

The reproducibility of the observations on the same subject at different times under similar conditions appeared to be quite good.

The mean head skin temperature for all subjects combined in the laboratory experiments was 36.8°C with helmet and 35.3°C without helmet. The difference was statistically significant ($p < 0.01$). The advantage of an open helmet was obvious, even under climatic chamber conditions with very little air movement, compared to a closed helmet. The relative humidity inside the open helmet was in the order of 55 to 70%, as against 100% in the closed helmet, and the head skin temperature was 0.3 to 0.5°C lower with the open helmet. The mean difference was statistically significant at a level of $p < 0.001$.

The results of the psychomotor performance tests in the laboratory showed a statistically significant improvement in reaction time in the simple-choice reaction time test after exercise vs. before exercise with ($p < 0.001$), and without ($p < 0.01$) helmet. This was presumably a result of the increased head temperature. In the case of the complex choice reaction time test, however, there was a statistically significant increase in the number of errors after exercise when using a helmet vs. not using helmets ($p < 0.05$).

The results of the outdoor bicycle rides performed by four of the six subjects, with and without helmet, showed that the ambient heat stress was low or moderate, and that the relative humidity inside the helmet was markedly lower than in the climatic chamber experiments (60 to 80 vs. 100% in the case of the closed helmet). Consequently, the skin temperature, including the head skin temperature was considerably lower than was the case in the laboratory.

The main finding during the outdoor rides was that the head skin temperature was consistently higher than the thigh skin temperature. An example of this is shown in Figure 5-20, where the head skin temperature was close to 35°C vs. 25°C at the thigh at the end of the ride.

The performance of one of the subjects is shown in Figures 5-21. Here again, the effect of the helmet on the head temperature is marked, as is the difference between head and leg temperature (Figure 5-21a and b).

It is quite clear that bicycling may lead to a hot head, whether using a helmet or not. But the head skin temperature may be significantly higher with a helmet than without, while an open helmet is somewhere in between.

The psychomotor tests peformed before and after exercise in controlled climatic chamber tests revealed a significant improvement in reaction time, concomitant with an increase in head temperature. This finding suggests an increase in speed of neurophysiological mechanisms underlying the mental alertness the speed of reaction. This may not be surprising in view of the fact that both cellular and metabolic reactions as well as nervous conduction rates are temperature dependent (for references, see Åstrand and Rodahl, 1986).

This, however, does not mean that the subject's ability to make the correct decisions is improved, rather it was to the contrary. As revealed by the complex psychomotor test, the number of errors increased when the subject was using a helmet and his head temperature was very high. These findings suggest that a moderate physical warming-up exercise may not only improve the individual's physical performance capacity, but also enhance cellular components of mental function, as well. However, excessive heating of the head may cause a deterioration in mental performance. Our findings do indicate that an open helmet is superior to a closed helmet in terms of preventing excessive heat stress of the head.

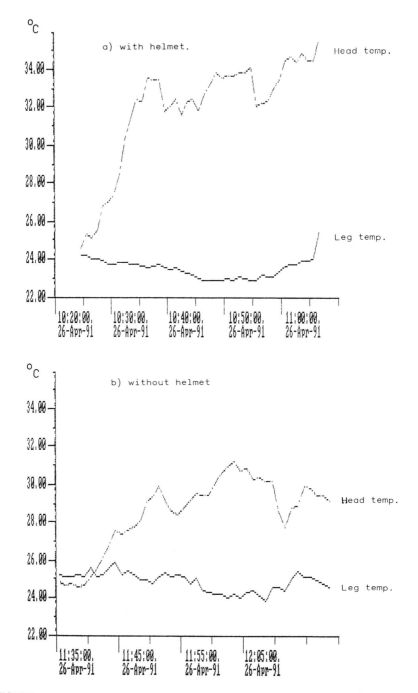

FIGURE 5-21. The effect of using a closed bicycle helmet on the forehead temperature in a subject during a 60-min outdoor bicycle ride. (Rodahl et al., The Fifth International Conference on Environmental Ergonomics, Maastricht, Nov. 2-6, 1992e. With permission.)

CHAPTER 6

The Sensing of Humidity and Dust

In a Nordic country such as Norway, during the winter, the relative humidity of the ambient air tends to be quite low due to the cold. This may cause dryness of the respiratory tract, contributing to the development of upper respiratory infections, especially when the relative humidity of the ambient air is below 30% for prolonged periods.

In the course of my many years of cold weather physiology engagement, I have been constantly looking for a practical way of solving this problem by being able to increase the relative humidity in the internal environment at home and at work. I was always told by leading authorities in the field that this was impossible until I came across a small, portable, ultrasound air humidifier named ADAX, Type LF6 (produced by Adax Factories, Ltd., Svelvik, Norway).

In a matter of minutes, this humidifier was capable of increasing the relative humidity in a bedroom by some 10%, as is evident from Figure 6-1. In this case, the Squirrel logger with two relative humidity sensors 20 cm apart was placed 1.25 m away from the humidifier and 50 cm above the floor. This is merely another example of the usefulness of being able to continuously log pertinent parameters with simple loggers, such as the Squirrel, in order to assess the relationship between cause and effect. This is done in an endeavor to improve the environment surrounding us at home and at work.

Another such common environmental problem affecting our health and well-being is airborne dust, especially respirable dust, i.e., dust particles in the air which are smaller than 10 μm. Continuous logging of such airborne dust concentrations are of considerable practical importance, both at home and at work, especially in the case of asthmatic or allergic individuals, and as a basis for constructive measures in order to improve the environment.

My own personal experience is limited to two types of aerosol monitors which I have used as sensors in combination with the Squirrel logger: the HAM (Handheld Aerosol Monitor, produced by Casella London Ltd.) and the Miniram (Miniature Real-Time Aerosol Monitor, produced by GCA Corp., Bedford, MA, U.S.A.).

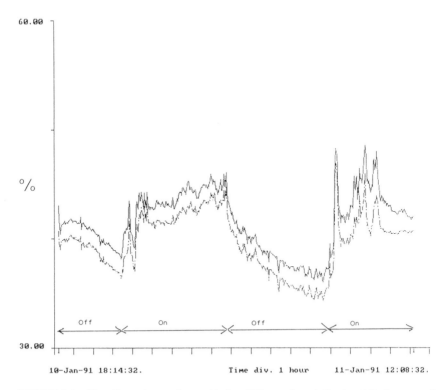

FIGURE 6-1. The effect of a small, portable humidifier on the relative humidity (in percent) in a bedroom. The two tracings are the results of two relative humidity sensors, 20 cm apart, 1.25 m away from the humidifier, and 50 cm above the floor.

The advantage with small, personal aerosol sensors, such as the Miniram, is that they can be carried on the individual operator to sense the aerosol concentrations where he or she is working even if the work is not stationary. The fact that it is compatible with an ambulatory logger such as the Squirrel with an extensive memory capacity makes it possible to visualize the aerosol concentrations from second to second. Combined with a detailed activity log, this will enable the investigator to find the course of the observed changes in the aerosol concentration. This could not be done with the traditional gravimetric methods, based on the collection of the dust on filters and subsequent weighing and calculation of the concentration on the basis of the measured air volume passing the filter. As a rule, this required sample-collection periods of several hours, which excluded any possible cause and effect study in the case of rapidly changing conditions.

Both of the above-mentioned aerosol sensors rely on ambient air movement to introduce airborne particles into the sensing chambers. This eliminates the need for a pump, and it's associated power supply which reduces the weight of the instrument considerably. They employ the technology of light scattering as a detection principle.

The HAM is, as the name indicates, a hand-carried instrument, which makes it less suitable than the Miniram for continuous ambulatory logging over an entire work shift without interfering with the operator's work procedures. The Miniram, on the other hand, can be pinned to the worker's coat.

In the compact Miniram monitor, the principle measurement is based on the detection of electromagnetic radiation. The scatter of the emitted radiation, caused by the aerosol particles, are recorded by a detector supplied with an amplifier. An optical interference filter is incorporated in the detector system in order to filter any signal which has a wave length different from that of the source. For a full description of this battery-powered light-scattering particle detection system, see Lilienfeld and Stern (1982). The Miniram is supplied with two different types of calibration kits, based on Arizona Road Dust. In addition, a systematic comparison with conventional gravimetric methods, based on simultaneous recording, is to be recommended. The Miniram sensor measures $10 \times 10 \times 5$ cm and weighs only 450 g.

Chung and Vaughan (1989) have tested the Miniram in a laboratory calm air chamber against a range of airborne dusts. The performance of the instrument was compared with that of a standard respirable dust sampler. They concluded that, after calibration for a given type of dust, the instrument could provide good estimates of the respirable dust mass concentration in the test chamber.

The Miniram–Squirrel combination was tested and compared with the conventional gravimetric aerosol measuring method with 8-hour sampling periods in an aluminum-producing plant in the north of Norway by Nes (personal communication) in 1990.

In an initial series of 12 parallel measurements, the Miniram readings, as originally calibrated by the factory standard, were compared with the results of the conventional gravimetric method. The mean value from the Miniram was 0.85 mg/m^3 vs. 1.53 mg/m^3 by the conventional method. The Miniram calibration was adjusted accordingly, and five new parallel tests were done. The results were then 0.93 mg/m^3 on the average on the Miniram vs. 1.14 mg/m^3 by the traditional method.

The results of this logging of the airborne dust concentration from the personal Miniram–Squirrel logger combination are shown in Figure 6-2. It should be emphasized that by far the greatest value of this continuous logging of parameters such as dust is the time-related picture it provides of the occurrence of the brief periods of increased dust levels, and possible causes for this occurrence. In this connection, it is the relative values that are of interest, and not so much the absolute values. If accuracy is essential, however, it is more a matter of devoting time to an accurate and carefully controlled calibration of the sensor with the use of the particular kind of dust in question.

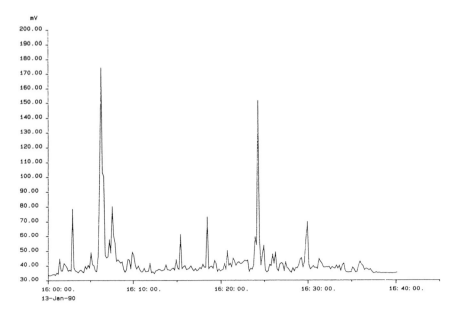

FIGURE 6-2. An example of the Miniram–Squirrel combination for the continuous logging of the airborne dust concentration in an aluminum-production plant (Nes, personal communication, 1990).

CHAPTER 7

The Logging of Carbon Monoxide (CO) Exposure

GENERAL CONSIDERATIONS

Toxic gases, such as CO, SO_2, H_2S, etc., represent an important aspect of air pollution as a consequence of modern industrial activity, automobile traffic, power plants, etc. In the past, this has mainly been an area of responsibility allotted to the industrial hygienists, but the problem of the pulmonary uptake of these noxious gases is also a matter of interest for the work physiologist who has a special competence in the field of pulmonary physiology. This has been the case because the uptake of these gases increases with increasing pulmonary ventilation as a result of increasing work rate. There is, therefore, reason to expect a much closer collaboration between the occupational hygienist, the work physiologist, and the plant physician in the solution of occupational health problems than has been the case in the past.

CARBON MONOXIDE CONCENTRATIONS IN THE AMBIENT AIR

Carbon monoxide is produced in any kind of incomplete combustion. This occurs in many industrial operations and it occurs in a burning cigarette. The toxic effect of the carbon monoxide is mainly due to its choking effect on vital cells, which, in extreme cases, may be lethal to the organism. The real problem is that carbon monoxide combines with hemoglobin in the blood with an affinity which is more than 200 times greater than that of oxygen. Since it is the hemoglobin which carries the oxygen to the different cells and tissues of the body, the amount of CO in the uptake replaces an equivalent amount of oxygen. Thus, the uptake of CO is at the expense of oxygen, hence a choking effect. And since the blood is so keen on picking up CO, it also hangs onto it that much longer. This means that habitual smokers have a tendency to accumulate greater amounts of CO, as carbon monoxyhemoglobin (COHb), at the expense of a corresponding amount of oxygen and to reduce endurance correspondingly. This is the reason top athletes in endurance events do not smoke. The smokers who inhale are especially apt to accumulate COHb, but even passive smokers are affected.

The amount of CO uptake distributed within the body depends on several factors. It depends on the CO content in the ambient air, which ideally is negligible, but which in extreme cases may reach concentrations of several hundred parts per million. The CO uptake also depends on the length of time the individual is exposed to the CO-contaminated air. Finally the CO uptake depends on the amount of CO-containing air which is breathed into the subject's lungs and thus, in contact with the blood and attached to the hemoglobin molecule. This exchange of air in the lungs is known as pulmonary ventilation. At rest, the pulmonary ventilation of an average individual may be less than 10 L/min. At maximal effort, it may be in excess of 100 L/min in untrained individuals and up to 200 L in trained athletes. (For further details: Åstrand and Rodahl, 1986; Rodahl, K., 1989). There is a linear relationship between work rate and pulmonary ventilation: the heavier the work rate, the greater the pulmonary ventilation. Similarly, the heart rate increases linearly with work rate. Therefore, heart rate may be used indirectly as an index of pulmonary ventilation.

The combined effect of the CO content of the inhaled air, duration of the exposure, and pulmonary ventilation will determine how much CO is accumulated in the blood as COHb. In the final analysis, this is the parameter that matters.

In the past, stationary systems have been used for the monitoring and control of the CO content of the ambient air in CO-exposed places of work. The disadvantage with these permanently installed, stationary systems is that they do not necessarily record the CO concentration at the locations where the operators work. The portable battery-operated CO meters, on the other hand, may measure the CO concentration at the location where the operators are actually working, but the concentrations may fluctuate so rapidly that it may be difficult to decide which reading to use as a representative figure. The advantage with the combination of ambient gas sensors and electronic loggers capable of storing the recordings in their electronic memories is that it provides for a more dynamic visualization of the CO exposures from one moment to the next.

However, the CO concentration in the ambient air is one thing, but the amount of CO uptake by the blood is of far more vital importance to the person involved, as already mentioned. Most of the existing tables showing the relationship between CO concentration in the inhaled air and the amount of carbon monoxyhemoglobin (COHb) in the blood at different lengths of exposure refer to individuals at rest. In view of the fact that pulmonary ventilation may increase by a factor of 10 or more when changing from rest to work, the uptake of CO by the blood may be similarly increased. For this reason, it is necessary to take the level of pulmonary ventilation into account when considering CO uptake in an individual working in a CO-containing atmosphere. This can be done by using one of the available multichannel loggers, such as the Squirrel, to record the CO concentration in the ambient air. At the same time, it can record the heart rate as an indication of work rate and, hence, pulmonary ventilation. Or it can record pulmonary ventilation directly with the aid of an appropriate air-flow sensor. Finally, the COHb percentage saturation can be determined by a Squirrel-connected Bedfont CO monitor. Thus, the relationship between the CO concentration in the respirable air and the CO uptake in the blood (at different work rates and different lengths of exposure) can be established. This is based on the well-established method of assessing COHb levels by measuring the CO concentration in the expired air at the end of expiration and after breath holding—as used in the Bedfont EC50 analyzer, produced by Bedfont Technical Instruments, Ltd., Sittingbourne, Kent, England (Jarvis et al., 1986; Irving et al., 1988).

COMPARISON OF THREE AVAILABLE CO SENSORS

The merits of three commercially available CO sensors in combination with the Squirrel logger were compared in a study aimed at measuring the carbon monoxide content of the ambient air in a typical silicon carbide (SiC) production plant in the south of Norway (Rodahl, K., 1989). They were the Sabre CO analyzer (produced by Sabre Gas Detection, Ltd., Guildford, Surrey, England); The Dräger Comapac CO analyzer (produced by Drägerwerk, Lybeck, Germany); and the Compur Monitox CO analyzer (produced by Compur Electronic, München, Germany). The specific cells of these sensors work on an electrochemical exchange principle. The ambient air diffuses through the membrane of the sensor cell. As the result of an electrochemical reaction, an electrical voltage is created, proportional to the reaction and to the partial pressure of the gases in the ambient air.

All three sensors were calibrated against known test gases containing 50 and 121 ppm CO in air. In the test gas, there was a straight line relationship between the concentration of CO (in parts per million) and the millivolts reading on the Squirrel (Figure 7-1), which was quite similar in all three sensors. A recalibration several weeks later showed almost identical results, indicating a high degree of stability in all three sensors.

All three sensors were used to measure the CO concentration of the air at different locations in the SiC production plant. They were plugged into the same Squirrel logger, so that the readings from the three sensors could be compared simultaneously with the readings of a conventional Commonwarn CO meter.

Detailed records of CO concentrations at different locations in the SiC-producing plant—both by permanently installed recording instruments and by ambulatory CO meters (Dräger Commonwarn)—have been kept over a number of years. From these records, it appears that the average concentrations of CO in the passage between the two rows of ovens may vary roughly between 20 and 80 ppm. In the adjacent hall, where the finished SiC was sorted, the levels were somewhat lower and more stable, around 30 and 40 ppm.

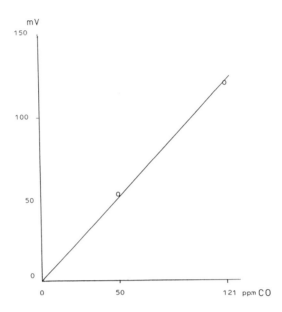

FIGURE 7-1. The straight-line relationship between the concentration of CO in the test gases and the millivolt reading on the Squirrel meter/logger, using a Compur CO sensor.

In the Production Hall of the plant, the conventional Dräger Commonwarn battery-operated, nonrecording CO meter showed extremely rapid and very marked fluctuations in the CO content of the ambient air. At times, the CO concentration might fluctuate as much as 50 ppm in a matter of seconds (Figure 7-2). This makes any kind of recording by a mechanical indicating meter difficult. This problem may be augmented by human factors, including the problem of deciding what meter reading to record. This is another reason why a portable, ambulatory logger system may be preferable, especially when the logger is set to record every second or every few seconds.

The CO concentration in the ambient air was measured 1 m above the floor at a number of specific locations in the Production Hall. In the Sorting Hall, the ambient air was measured every 5 seconds by the Dräger Commonwarn. Each location compared the Dräger measurements with the three Squirrel-connected sensors which logged every second. As long as the ambient air was fairly stable, as was the case in the Sorting Hall, the agreement between the four sensors was fairly good, i.e., within 4 ppm.

There were considerable differences, however, when the air was turbulent, as in the passage between the ovens in the Production Hall. The turbulence was due to draft or moving hot objects in the vicinity. The difference was most probably due to the fact that the Dräger Commonwarn reacted much faster because it, unlike the other sensors, is actively pumping the air across its sensor cell.

THE LOGGING OF CARBON MONOXIDE (CO) EXPOSURE

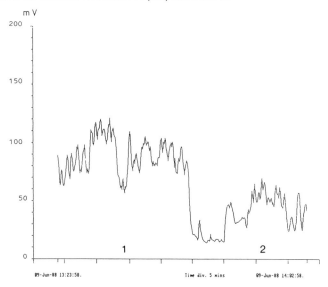

FIGURE 7-2. CO concentration in the ambient air in a SiC production plant, in millivolts, recorded by the Compur sensor plugged into the Squirrel meter/logger (1 mV is roughly equal to 1 ppm CO). The tracing shows extremely rapid and marked fluctuations in the CO content of the ambient air, both in (1) the production hall, and (2) the sorting hall.

It is thus evident that at the lower and fairly stable CO concentrations, all three Squirrel-attached test sensors were usable for the practical purpose of assessing ambient CO concentrations. The main difficulty, as already addressed, was the very rapid and marked fluctuations in the CO content of the ambient air, especially in the Production Hall (see Figure 7-2).

In an attempt to obtain stable samples for comparative measurements, samples of ambient air were collected at two typical locations in the plant. The air was collected in large airtight plastic bags, with the aid of a pump. This trapped air was then thoroughly mixed and analyzed in the laboratory by the standard Dräger Commonwarn CO meter and the Ecolyser CO meter. These readings were then compared with the readings of the three Squirrel-attached test sensors. To secure an even flow, a constant weight was placed on top of the bag. The results showed a reasonable degree of agreement. At the 130-ppm level, the readings with the Compur and the Commonwarn sensors were identical. The Dräger CO sensor showed values 6 ppm higher and the Sabre 15 ppm higher than the Commonwarn reference values. At the 40 ppm level, the difference between the standard reference sensor and the three trial ones varied from 3 to 7 ppm.

It is thus evident that the CO sensors that are already available, combined with a logger such as the Squirrel, can, for all practical purposes, be used for the routine recording of CO concentrations in the ambient air in a variety of industrial work places. They can also be used for the purpose of assessing CO concentrations in the ambient air in areas with heavy automobile traffic, in tunnels, etc.

ASSESSMENT OF THE ACTUAL CO UPTAKE BY ESTIMATING THE COHb CONTENT OF THE BLOOD WITH THE BEDFONT CO MONITOR

The percentage COHb saturation of the blood is routinely directly assessed by the analysis of a venous blood sample and using conventional methods. It is well established, however, that alveolar breath analysis may be used for a fairly accurate estimation of the percent COHb saturation in healthy individuals, provided the ambient CO concentration is near zero (Peterson, 1970; Rea et al., 1973; Rees et al., 1980; Jarvis et al., 1980; Wald et al., 1981).

The Bedfont Carbon Monoxide Monitor, Model EC50, is an inexpensive, easy-to-operate instrument. It is specifically designed to measure CO concentrations in the expired air, as an indication of COHb content expressed as percent saturation. Jarvis et al. (1986) have compared the results obtained by this monitor with measured blood COHb concentrations in 72 subjects. The subjects had COHb concentrations which ranged from 0.3 to 12%. The expired-air CO concentrations measured by the monitor correlated closely with the COHb concentrations. The relationship was linear over this range. As an approximate guide, the percentage of COHb saturation could be obtained by dividing the expired air reading by 6. Irving et al. (1988) have compared the same instrument with the Ecolyzer in 138 normal subjects and confirmed the findings of Jarvis et al. (1986).

In the above-mentioned studies, the emphasis was on the application of the Bedfont monitor as an instrument to estimate tobacco smoking status, and as a motivational device for those attempting to stop smoking. Another interest in the monitor stems from the need to protect occupational health personnel in CO-exposed industry. This could be a reliable instrument for assessing COHb saturation in their workers following exposure on their jobs. For this reason, the Bedfont CO monitor was connected with the 1201 Squirrel logger, supplied with an event marker for the loggings of individual readings at specific times. This combination was used in a pilot study performed at the Pulmonary Department of Haukeland Hospital in Bergen, Norway (Rodahl, K., 1989). The instrument combination was used to record CO concentrations in parts per million in the expired air from two subjects who were nonsmokers, from one subject who said that she only smoked 10 to 15 cigarettes per day, and from one subject who said that he smoked in excess of 25 cigarettes per day.

The study was performed in a laboratory where the CO content of the ambient air was negligable. One subject at a time was tested while seated in a chair. Prior to the actual test, which was repeated twice, the subject was instructed in the use of the monitor and allowed to practice. A venous blood sample was drawn from the cubital vein for COHb analysis by a OSM3 Hemoximeter (Zijlstra et al., 1988). The expired air was measured by the CO monitor. The procedure for the expired-air CO measurements was as follows: the subject was asked to exhale, then to inhale as deeply as possible, hold the breath for 20 seconds (recorded by stopwatch), and then to exhale into the Bedfont mouthpiece forcefully and as

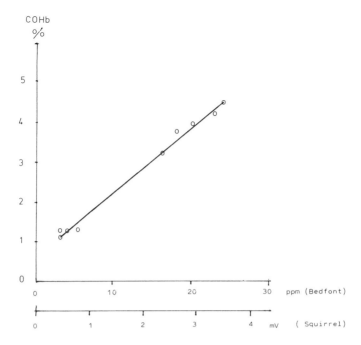

FIGURE 7-3. The relationship between CO content in expired air (read in parts per million on the Belfont CO monitor and in millivolts on the Squirrel meter/logger) and the photometrically measured percentage COHb saturation.

completely as possible, pressing the elbows against the chest. The readings on the Bedfont monitor (ppm) were continuously monitored, and compared with the millivolt readings on the Squirrel.

The results showed a straight line relationship between parts per million in the expired air and the photometrically measured percentage COHb saturation. There was also a linear relationship between the Squirrel readings in millivolts and the measured percentage COHb saturation (Figure 7-3). The difference between the estimated COHb saturation (based on the expired-air CO and according to the graph supplied with the instrument) and the actually measured percentage COHb saturation in the blood sample varied from 0.3 to 0.7%. On the average, it appears that the Bedfont CO monitor underestimated the COHb saturation by 0.6%.

This pilot study was followed by a field study at a typical CO-exposed work place in which the Bedfont CO monitor was tested in combination with sensors for the assessment of CO content of the ambient air. A heart rate recorder, used to determine work rate, was plugged into the same Squirrel logger. The results of the Bedfont-predicted percentage of COHb saturation were compared with the results of the analysis of COHb percent saturation in blood samples from the same subjects taken at the same time. For further details, see Rodahl, K., (1989).

The subjects participating in the field study were exposed to different work loads for 15 min using a cycle ergometer to ensure constant and repeatable work loads. In view of the fact that heavy tobacco smokers have as much as 5 to 10% COHb saturation due to the CO uptake from the tobacco smoke alone, both smokers and nonsmokers were used as subjects for this field study. The six subjects who volunteered for the study were all employees of the company, working in the CO atmosphere of the production plant. Three of them were smokers and three were nonsmokers. They reported at 9 a.m. at the laboratory where the CO content of the ambient air was close to zero. They had not been exposed to the CO-containing atmosphere of the SiC-production area earlier that day. They remained in the CO-free laboratory area until their turn came to take part in the test exposure. Prior to the CO exposure, blood samples were drawn for the analysis of the percentage of COHb saturation. The CO content of the alveolar air was then measured by the Bedfont CO monitor, as previously described, for the prediction of percent COHb saturation. The Compur CO sensor was attached to the subject's coat, close to his neck, and connected to the Squirrel logger carried by the subject, attached to his belt. The Squirrel logger was set to record the CO content once a minute. This was compared with stationary recording of the CO content of the ambient air by another Compur sensor attached to a separate Squirrel logger. The Compur sensor was placed about 1.5 m away from the cycle ergometer where the subject was to be situated during the test. A heart rate recorder was connected to the Squirrel logger which was carried by the subject. The heart rate recorder was mounted on the subject for the continuous logging of his work rate. The subject then walked out to the SiC-production plant, where each of subjects 1 through 5 spent 15 min in the Sorting Hall. Subject 6 went to the adjacent end of the Production Hall where the CO content of the ambient air was considerably higher. Subjects 1 through 3 were smokers, subjects 4 through 6 were nonsmokers. The CO content of the ambient air was continuously recorded and logged once every second by a stationary sensor. It was read once every tenth second on a Dräger CO meter placed next to the Compur sensor which was located at the level of the subject's head at a distance of 1.5 m.

The two Compur CO sensors used were calibrated prior to the test (50 mV = 50 ppm CO). The heart rate recorder had been checked against ECG recordings and manual pulse counts.

At the end of the 15-min exposure period, the subject returned to the laboratory where a blood sample was drawn for COHb saturation analysis. The alveolar CO-content was then assessed by the Bedfont CO monitor as previously described. In this case, the test was repeated three times, and the highest values were used.

The results of the stationary recording of the CO content of the ambient air once every second during the entire period of the tests showed great and extremely rapid fluctuations in the CO content of the ambient air surrounding the subject. The values for CO content, recorded and logged every second by the Compur sensor, and the values read by the Dräger meter once every 10 seconds, were very similar. In contrast to this, there were considerable differences between the Compur sensor values recorded once every second, and the values recorded once every minute from the Compur sensor carried on the subject, indicating that recordings once a minute may give an erroneous picture of the real CO-exposure level. (For further details see Rodahl, K., 1989).

The mean difference between predicted and actually measured COHb saturation in the blood was no more than 1% before, and 0.7% after the CO exposure. The difference was particularly small at the lower values, i.e., in the nonsmoking subjects.

From these results, it appears that, for practical purposes, the Bedfont–Squirrel method is quite applicable for predicting the CO-saturation level in the blood of workers in the CO-exposed industry. The technique is simple and can be used by the industry's own technicians without scientific training, and after a brief course of instruction. The interpretation of the results is also quite simple. Furthermore, it is possible to make a similar assessment of COHb saturation as that achieved by the Bedfont monitor by using any Squirrel-compatible CO sensor such as those previously mentioned (Sabre, Compur, Dräger, etc.). A proper mouthpiece and gas chamber placed over the sensor membrane connected to a Squirrel logger should be supplied. In this manner, the same CO sensor may be used both to log the CO content of the ambient air, and to assess the percentage COHb saturation in the exposed subject's blood.

By this logger system, it is also possible to measure and log the physical work rate of the CO-exposed individual. This permits a systematic analysis of the relative importance of this parameter. It is also technically possible to supplement the instrument assembly with sensors or probes which measure the pulmonary ventilation directly; for instance, by using flow meters compatible with the Squirrel logger.

By far the most striking observation made during this project was the important role played by the worker's smoking habits for the maintenance of a high COHb accumulation in the blood, both during work and leisure. The difference in the basic preexposure levels of COHb saturation between smokers and nonsmokers was much greater than the difference in COHb saturation before and after the CO exposure. One of the smokers had a COHb saturation of 8.5% before the CO exposure, while the greatest increase in COHb saturation observed during the CO exposure was only 1.2%. This was in a worker who was exposed to a mean CO concentration in the ambient air of 107 ppm for 15 min and subject to a work load of 600 kpm/min. It is thus evident that tobacco smoking may be more important than brief periods of industrial exposure to carbon monoxide concerning the CO accumulation in the blood.

It was also interesting to note that in all three smokers studied, the percentage COHb saturation was higher before than after CO exposure. This may possibly be due to the fact that they had to refrain from smoking during the entire period of the experiment, which lasted most of the day.

The role played by the level of the physical work performed by the subjects during the experiment is unclear. Nor is it clear whether or not the work loads used in this experiment were representative of the work loads to which these workers were exposed during their ordinary daily work in the Production and Sorting Halls. It should be noted, however, that the COHb saturation tests in two individuals who spent 10 min walking along the pathways between the two rows of ovens in the Production Hall showed an increase in the predicted percentage COHb saturation from 1.0 to 1.8 in the nonsmoker, and from 4.2 to 4.7 in a smoker. This indicates that even brief periods of exposure to the prevailing level of CO concentrations in the ambient air in some industrial plants may cause a considerable COHb accumulation.

The main purpose of this project was to contribute to the development and field testing of a method for an objective survey of the CO environment, combined with a noninvasive method of estimating the percentage COHb saturation of the blood as a meaningful expression of the physiological consequences of the CO exposure. While in essence, this objective was attained, the project also brought up a number of practical and basic questions which may be answered by the use of the applied method. This includes a more systematic and detailed mapping of the CO exposure and the effect this may have on the COHb saturation of the workers at the different work places. It would be equally interesting to learn how long it takes for the increased COHb saturation to return to its initial level following the exposure. It would also be of interest to examine further the effect of the work rate on the CO uptake, and to what extent the level of physical fitness may influence this relationship.

CHAPTER 8

The Logging of Sulfur Dioxide (SO_2) Exposure

The effect of SO_2 exposure on the health of the exposed workers in certain types of Norwegian industry has been a subject of increasing attention in recent years. From all indications, however, there is still no convincing evidence of a clear relationship between industrial SO_2 exposure and the development of any chronic pathological conditions. It has been claimed that SO_2 exposure (above certain concentrations) may elicit attacks of the so-called "pot-room asthma" in the aluminum industry. So far, however, this has not been confirmed scientifically. On the other hand, it is a common experience that SO_2 exposure to concentrations above 2 ppm does cause mucous membrane irritation in the respiratory tract when inhaled. The moisture of the mucous membrane in the respiratory tract may cause the SO_2 to be converted to H_2SO_3 or H_2SO_4. In spite of this, controlled laboratory experiments have shown that inhalation of air containing 2.0 ppm SO_2 does not affect the pulmonary ventilatory functions in healthy individuals (Bedi and Horvath, 1989).

So far, the recommended upper limit for the SO_2 concentration of ambient air in most industrialized countries has been 2 ppm. In Norway in 1991, regarding the contemplated lowering of this limit to 1 or even 0.5 ppm, we were asked to conduct a survey of the SO_2 concentration in two of the production halls in a typical silicon carbide (SiC) production plant in the south of Norway (Guthe et al., 1991). The plant produces SiC by placing the raw material, which consists of coke and quartz in long heaps known as production beds, around electrodes which heat the raw material and cause it to react, thereby forming the SiC. The highest ambient SO_2 concentrations are associated with the pulling down of the production beds. Previously, the SO_2 concentrations in the ambient air in the SiC production halls (and their variations during the SiC production process) had only been recorded occasionally. It was evident, however, that SO_2 usually appears together with carbon monoxide (CO) in the working environment, and has the advantage over CO—as far as the operators are concerned—that its presence can be smelled.

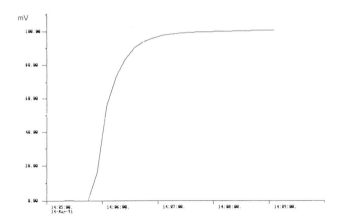

FIGURE 8-1. The calibration curve for one of the Compur SO_2 sensors used in the study. (1 mV = approximately 0.10 ppm.) (From Guthe et al., En orienterende logging av svoveldioksyd innholdet i arbeidsatmosfæren ved Arendal Svelte verk, 11-13 mars 1991. Upublisert rapport, 1991. With permission.)

The measurement of the SO_2 concentration in the air was done with Compur SO_2 sensors clipped to the chest pocket of the subject's jacket and connected to a Squirrel logger, which was fastened to the subject's belt. Two Compur sensors were used, allowing us to log the SO_2 exposure of two different subjects at the same time. Three of the regular operators in the two production halls volunteered to serve as subjects in the study, one of them on two different shifts.

The sensors were calibrated before the study, and recalibrated immediately after the study, against a test gas containing 9.6 ppm SO_2. A Squirrel deflection of 1 mV equalled approximately 0.10 ppm (Figure 8-1).

The readings were stored in the electronic memory of the Squirrel logger once every 10 seconds. At the end of the observation period, the stored data were transferred to a portable PC. They were visualized in the form of graphs and in the form of tables showing the observed values in relation to the recommended maximal values. They were printed on a conventional PC printer.

Two men and one woman volunteered for the study, which involved two morning shifts and one afternoon shift in one of the production halls, and one morning shift in another production hall. The study included more or less all phases of the SiC production process, from the building to pulling down the SiC production bed.

In addition to this SO_2 logging by a sensor carried on the person, we made a systematic logging of the ambient SO_2 concentrations in the different production areas in the two production halls included in the study as well as in one of the crane cabins in one of the production halls. The results of the Compur–Squirrel logging were compared at certain intervals with the readings on a manual Dräger spot recorder.

The results of the above-mentioned ambulatory survey of the SO_2 concentrations in the working atmosphere in the two production halls are shown in Figure 8-2.

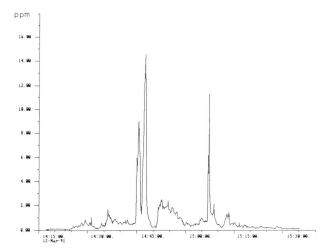

FIGURE 8-2. Ambient SO_2 concentrations (ppm) in the working environment, logged by an ambulatory Compur sensor–Squirrel logger combination, carried by a person walking through two SiC production halls. The peak values were recorded in the vicinity of the active SiC beds. (From Guthe et al., En orienterende logging av svoveldioksyd innholdet i arbeidsatmosfæren ved Arendal Svelte verk, 11-13 mars 1991. Upublisert rapport, 1991. With permission.)

From this figure, it appears that the SO_2 concentration in the ambient air varies greatly. The highest values were recorded in the vicinity of the glowing hot mixture of coke and quartz in the active SiC beds. The concentrations also varied greatly from one moment to the next in the same place, as a consequence of draft and turbulent air movement in the hall. The mean SO_2 concentration during the entire observation period of 70 min was 0.85 ppm, ranging from 0 to 15 ppm. The SO_2 concentration exceeded 0.5 ppm 40% of the time in the production halls. It exceeded 1 ppm in 19% of the observation period, and was higher than 2 ppm 6% of the time.

A comparison of the observed readings on the Squirrel logger and the readings on the Dräger manual spot–pump meter showed a reasonable degree of agreement considering the limitation of the Dräger meter. During the time it took to pump air 10 times through the Dräger meter, the displayed readings on the Squirrel logger fluctuated from second to second by an amount several times the initial starting value.

Subject 1, who was working a morning shift in one of the production halls (Figure 8-3), was exposed to short periods of considerable SO_2 exposure. The SO_2 exposure had values with spikes to 8 ppm. These matched the five times he moved in and out of the space between the active beds to assist the crane operator in picking up sections of the forms supporting the beds (in Figure 8-3). The mean SO_2 exposure during the entire observation period was 0.6 ppm, varying from 0 to 18 ppm.

The same operator was studied a second time when again working a morning shift in the same production hall. This time he was exposed to a mean SO_2 concentration of 0.8 ppm during the entire work shift, with variations from 0 to

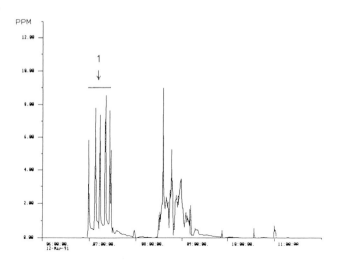

FIGURE 8-3. SO$_2$ exposure in a worker in a SiC production plant. He was exposed to brief periods of considerable SO$_2$ concentrations the five times he moved in and out of the space between the active beds to assist the crane operator in picking up sections of the forms supporting the beds (1). (From Guthe et al., En orienterende logging av svoveldioksyd innholdet i arbeidsatmosfæren ved Arendal Svelte verk, 11-13 mars 1991. Upublisert rapport, 1991. With permission.)

12 ppm. Here again the highest CO$_2$ concentrations were encountered when working near the active SiC beds.

Operator 2, working an afternoon shift in one of the production halls, was exposed to a mean SO$_2$ concentration of 1 ppm during the shift, with variations from 0 to 10 ppm (Figure 8-4). The SO$_2$ concentration to which she was exposed exceeded 2 ppm for a total of 30 min during the shift. The highest SO$_2$ concentrations were encountered when working close to active beds and when tearing down the beds where the SiC producing process had been completed.

The third operator was studied during a morning shift in one of the production halls engaged in a variety of operations quite typical of this type of industry. The mean SO$_2$ exposure for the entire shift amounted to 0.4 ppm, varying from 0 to 16 ppm.

Early in the morning this operator joined Operator 1 at a point near one of the active SiC beds where the SO$_2$ concentration was known to be quite high. The operator did this for the purpose of comparing the two Compur SO$_2$ sensors when exposed to the ambient SO$_2$ concentrations in the same place. The mean value from one of the sensors was 3.7 ppm (range 0.5 to 16.0, standard deviation 3.8) vs. 3.4 ppm (range 0.5 to 12.0, standard deviation 3.6) from the other. This is indeed a reasonable agreement, considering the vast fluctuations in the ambient SO$_2$ concentration at the sample area. The fact that the values were recorded in the Squirrel memory only once every 10 seconds made it likely that the two loggers might not be recording the same spikes since the exact time of storing the data in the loggers' memory might not be the same.

THE LOGGING OF SULFUR DIOXIDE (SO$_2$) EXPOSURE

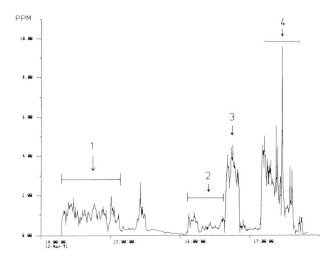

FIGURE 8-4. SO$_2$ exposure of a female operator in one of the SiC production halls: (1) preparing a SiC bed; (2) raking; (3) working between active SiC beds; (4) tearing down a completed SiC bed. (From Guthe et al., En orienterende logging av svoveldioksyd innholdet i arbeidsatmosfæren ved Arendal Svelte verk, 11-13 mars 1991. Upublisert rapport, 1991. With permission.)

Table 1 SO$_2$ Concentration in the Working Atmosphere

Operator no.	Above 0.5 ppm	Above 1.0 ppm	Above 2.0 ppm
1(a)	26%	17%	9%
1(b)	40%	21%	10%
2	53%	23%	14%
3	21%	19%	2%

From this survey it is evident that the SO$_2$ concentration in the ambient air varied greatly in the SiC production halls which were examined. The closer to the active burning SiC beds, the higher the concentration. Using the standard Squirrel softwear program, the length of time was calculated. Time was expressed as percentage of the entire observation period in which the SO$_2$ concentration in the working atmosphere exceeded 0.5, 1.0, and 2.0 ppm. The observation period was, for all practical purposes, equal to the entire work shift. The results are presented in Table 1.

In this particular SiC producing plant and on the basis of these results, it would appear that a lowering of the permissable upper limit of SO$_2$ concentrations in the working atmosphere to 0.5 ppm, in all probability would mean that the plant would have to close.

It should be kept in mind, however, that the above-presented values are limited to (1) the existing conditions in the production halls examined in this particular SiC plant at the time of the study; and (2) under the prevailing conditions at the time, including the level of production, the number of active beds, and their relative location. Whether or not these findings may be taken as representative of similar plants with similar production levels can only be determined by similar loggings at the plants in question. This is due to a number of conditions, especially ventilation, air circulation, etc., which depend on the construction and other aspects of the production hall, and which do affect the SO_2 concentration in the working atmosphere.

The survey does show, however, that the logging method used for the purpose of continuously visualizing the SO_2 exposure may also be used to arrive at working routines which result in the lowest SO_2 exposures. This is rather convincingly demonstrated in Figure 8-3, showing the spiky SO_2 concentrations to which the operator is exposed every time he enters the space between the active SiC producing beds to assist the crane operator in his operations.

Another advantage with this logging system is that it allows the logging of several relevant or interacting parameters simultaneously in addition to SO_2, such as CO and dust.

CHAPTER 9

Continuous Logging of Hydrogen Sulfide (H_2S)

The concern for hydrogen sulfide (H_2S) as an air pollutant is primarily due to its toxic effect, apart from its unpleasant odor and its reactivity with metals and metal salts (Natusch and Slatt, 1978). Sewage is one of the common sources of H_2S contamination (Kangas et al., 1986). In a study involving a number of Norwegian sewage plants (Heldal et al., 1992), exposure to short periods of excessive concentrations of H_2S appeared to be associated with symptoms such as sore eyes, fatigue, and reduced ability to concentrate.

There are several electrochemical H_2S sensors available on the market for ambulatory logging of individual H_2S exposure. One such sensor is the Compur 4100 Monitox H_2S sensor produced by Compur–Electronic GmbH, which can be used in combination with the ambulatory Squirrel minilogger (produced by Grant Instruments, Ltd., Cambridge, England). The reliability and accuracy of such ambulatory H_2S sensors have been tested by Accorsi and Hure (1990).

This sensor–logger combination was used by Søstrand (1992) for the purpose of assessing the level of H_2S exposure in 23 sewage workers in the south of Norway. The ambulatory logging was carried out for a period of 4 hours each day for 2 days in each of the 23 subjects. The results showed that 6 of these 23 workers were exposed to H_2S. Although the time weighted average (TWA) was less than 1 ppm H_2S for all six workers, two of the operators in a pumping station were exposed to peak concentrations of 25 and 45 ppm during sludge washing and four of the workers in four different sewage plants were exposed to peak concentrations from 3 to 15 ppm of H_2S. These examples clearly illustrate the value of continuous ambulatory logging of personal exposure to noxious gases as an indication of the actual moment to moment exposure of the workers involved. In contrast, stationary average readings over prolonged periods of time, perhaps even including periods when there is no work occurring, may be the time when H_2S is most likely to be stirred into the air.

In a subsequent study, Søstrand (1992) recorded the ambient concentration of noxious gases, including H_2S, in 15 sewage plants and in two pumping stations in the south of Norway. In this case, a Neotronic GS 8620 electrochemical sensor

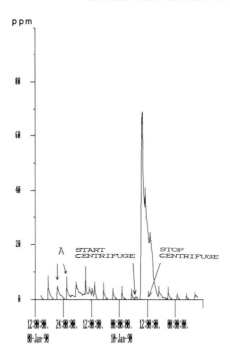

FIGURE 9-1. The changes in H_2S concentrations (parts per million) in a sewage plant, when the sewage was pumped once every 4 hours into a catch pit (A), and when the sludge was centrifuged for a period of several hours in a spin separator. (From Søstrand, P., Nat. Inst. Occup. Health, Oslo, Norway, unpublished, 1992.)

(produced by Bieler and Lang in Denmark) was used in combination with the Squirrel 1200-series logger for stationary recording. The sensor was placed 1.5 m above the floor, close to basins or sludge treatment units such as spin separators, sludge presses, etc. The logger was set to record at 2-min intervals in order to reveal short-lasting peak concentrations.

One of the sewage plants was a small but modern unit dimensioned to handle sewage from a community consisting of 2000 people. The sewage entered the plant through sewage pipes. After mechanical filtration, the sewage was treated with aluminum sulfate and the pH was adjusted to about six in a flocculating basin. Then it was transferred into settling basins where organic materials and aluminum phosphate were removed. From these basins, the sludge was pumped, once every 4 hours, into a catch pit which was placed in the middle of the hall. From this pit, the sludge was transferred into two sludge storage containers, where it was kept for a couple of weeks. It was then pumped with the aid of an eccentric screw jack into a spin separator, where it was centrifuged for a period of several hours for the purpose of separating solid matter from the liquid.

The small spikes of H_2S concentrations in the order of a couple of parts per million, occurring about every 4 hours (Figure 9-1, A), coincided with the pumping of the sludge from the settling basins into the catch pit. The very high, spiky

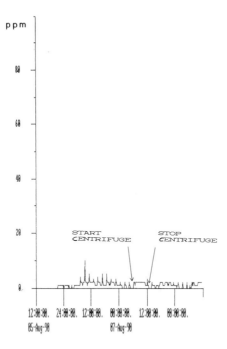

FIGURE 9-2. The effect of supplying the sludge storage containers in a sewage plant with a stirring mechanism which prevented the H_2S accumulation. (From Søstrand, P., Nat. Inst. Occup. Health, Oslo, Norway, unpublished, 1992.)

H_2S concentration exceeding 100 ppm (Figure 9-1) was associated with the transfer of the sludge from the storage containers into the spin separator and the centrifugation process.

On the basis of these findings, it became evident that the cause of the extremely high H_2S spike during the process of transferring the content from the storage containers into the spin separator for centrifugation was due to the release of H_2S from the settled sediment at the bottom of the storage containers. When this was remedied by supplying the storage containers with a stirring mechanism which prevented the H_2S accumulation, the spiky emission of H_2S into the ambient air disappeared (Figure 9-2).

The smaller H_2S emissions into the ambient air in the sewage plant every 4 hours, associated with the transfer of the sludge from the settling basins into the catch pit, were eliminated by installing a proper fresh-air ventilation system in the sewage plant (Figure 9-3). This is yet another example of the usefulness of being able to visualize pertinent parameters in order to determine the causes of the effects observed.

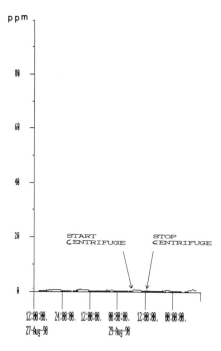

FIGURE 9-3. The effect of installing a proper air ventilation system in the sewage plant on the H_2S emissions into the ambient air every 4 hours, associated with the transfer of the sludge from the settling basins into the catch pit shown in Figures 9-1 and 9-2. (From Søstrand, P., Nat. Inst. Occup. Health, Oslo, Norway, unpublished, 1992.)

CHAPTER 10

A Simple Way of Assessing the Relative Concentrations of Chemical or Organic Vapors in the Working Atmosphere

The potential harmful effects of the numerous kinds of chemical and organic vapors in the working atmosphere on those who are exposed to them has been the subject of public discussion for a number of years. Special attention has been focused on some of the professional painters who claimed to suffer deterioration of mental functions as a consequence of prolonged exposure to the vapor from the solvents which they regularly use in their daily work. From these general discussions, one is left with the impression that much of it is based on assumptions rather than objective figures and facts.

With the new generation of applicable minisensor–logger combinations available, there is no longer any reason for not providing a clear-cut, on the spot, visualization of the magnitude of vapor present in the working atmosphere. The changes in their concentration with time and with the different work operations performed should also be provided as collected at the spot where the operators are actually working.

In the Fall of 1991, we had the opportunity of making a general, on-the-spot survey of the concentration and distribution of such vapors in a paint plant in Oslo producing varnish for the automobile industry (Rodahl et al., 1991b).

The sensor which was used in combination with the 1200-series Squirrel logger was the HNU Model PI-101 photo-ionizer (produced by HNU Systems Inc., Newton, MA, U.S.A.). This is a trace-gas analyzer capable of measuring the concentrations of a great variety of vapors in industrial atmospheres. It operates on the principle of photo-ionization for detection of the gas molecules. The sensor consists of a sealed ultraviolet light source emitting photons which ionize the vapor, particularly organic vapors. A chamber adjacent to the ultraviolet source contains a pair of electrodes. When a positive potential is applied to one electrode, the created field drives any ions formed by the absorption of ultraviolet light to the collector electrode where the current, which is proportional to the concentration, is measured. An example of two calibration curves showing the relative responses to toluene and

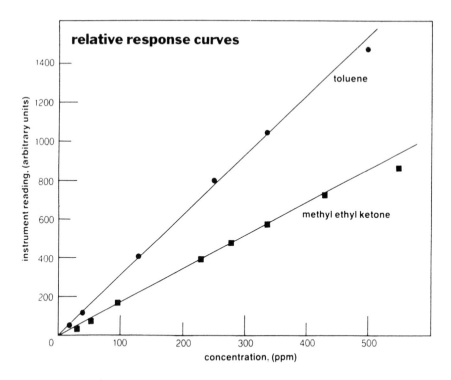

FIGURE 10-1. An example of two calibration curves, supplied by the producer of the HNU photo-ionizer, showing the relative responses to toluene and methyl ethyl ketone, at the same gain setting.

methyl ethyl ketone at the same gain setting, and supplied by the producer of the HNU photo-ionizer, is shown in Figure 10-1. Another example of the calibration of the sensor against a test gas containing 35 ± 5 ppm of N-butyl acetate in nitrogen is shown in Figure 10-2.

In the first survey, the HNU photo-ionizer, connected with a 12-bit 1200-series Squirrel logger, was carried by the investigator through two of the production rooms of the plant. The main paint production was taking place in a room (No. 2 in Figure 10-3) where the tip of the probe was held under the lid covering the production containers in which the different components of the particular paint under production were being mixed. As is evident from Figure 10-3, the overall concentration of the vapors in the ambient air in the two production rooms was quite different, but rather stable. In the case of the mixing room (room No. 2) where the main paint production was taking place, the vapor concentration was naturally much higher inside the actual paint containers than in the ambient air of the room itself ("a" and "b" in Figure 10-3).

In a subsequent survey in the same plant during the production of varnish for the automobile industry, a representative series of typical operations were assessed. This included preparations, mixing of the different ingredients, the automated tapping of the finished varnish into cans, and finally the cleaning of the

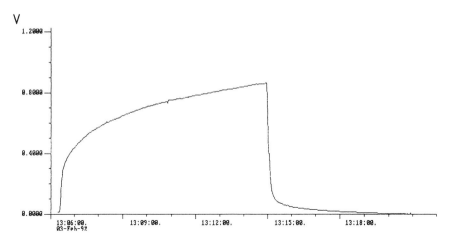

FIGURE 10-2. Calibration curve for N-butyl acetate, using the HNU photo-ionizer in combination with the Squirrel logger. 1 V is approximately 35 ppm of N-butyl acetate.

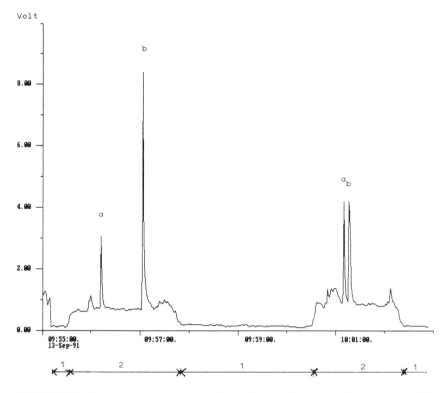

FIGURE 10-3. The vapor concentration in the ambient air in a paint plant, using a HNU photo-ionizer connected to a 12-bit Squirrel logger, carried through two of the production rooms (1 and 2) in a paint plant. (a and b) show the vapor concentration inside two paint-mixing containers.

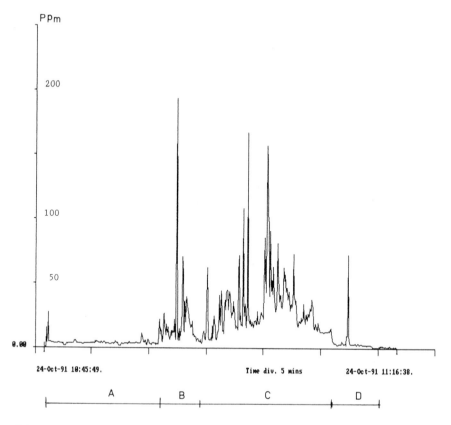

FIGURE 10-4. Vapor concentrations in the ambient air in a paint plant: (A) automated transfer of finished varnish into cans; (B) washing of varnish containers; (C) ambient vapor concentration in the mixing room; (D) vapor concentration above the floor when transferring varnish into a can.

empty containers after the mixing process was completed. The main exposure in one of the rooms was said to be white spirits, and a combination of acetone and butyl acetate in the other.

From Figure 10-4(A), it is observed that the automated transfer of the finished varnish through closed pipes into the cans was causing a very modest contamination of the working atmosphere, in the order of less than 5 ppm. The washing of the containers after they had been emptied was associated with a considerable increase in the vapor in the working atmosphere, both when using acetone and butyl acetate (B in Figure 10-4). The general ambient vapor concentration in the mixing room varied between some 20 and 35 ppm with short-lasting peaks roughly between 50 and 150 ppm when placing the tip of the sensor probe above the mixing containers (C in Figure 10-4). Finally, one of the workers had placed himself on a small box in the middle of the floor where there was a considerable amount of draft, while transferring some specially produced varnish into a 1-gallon can. This caused a brief spiky elevation of the vapor concentration (D in Figure 10-4).

From these few examples, it is obvious that the sensor–logger combination used here may be applied to perform a general on-the-spot mapping of the magnitude and location of vapor concentrations from different kinds of solvents used in paint production and similar kinds of industry. Since the industry in question does know exactly which solvents they use, the sensor can be calibrated against these solvent vapors, enabling a fairly accurate assessment of the contamination in the working atmosphere in general. The main advantage, however, with this combination of ambulatory sensor–logger system, is its use to determine the general concentration of the vapors in question, as well as their relative concentrations in different locations and, according to the working procedures, as a reference point for the evaluation of the effect of improvement in the working procedures or the working techniques.

A further example of the application of this sensor–logging system for the assessment of ambient vapor concentration is a survey made in a classroom for aircraft maintenance students. Here the students, in a classroom situation, were exposed to varying concentrations of vapors from different solvents used in the treatment of aircraft fuel tanks and the painting of aircrafts. From Figure 10-5 it is evident that the vapor concentrations in the classroom air at times may be considerable. It is also of interest to note the difference in the vapor concentration in the air under the hood, depending on whether or not the main door to the classroom was open or closed (Figure 10-6).

Because of its small size and weight, it is also possible to use this sensor–logger combination for the purpose of assessing the actual vapor exposure to which the aircraft maintenance workers are exposed, on the spot, while they are actually doing their maintenance work, even inside the fuel tanks and other enclosed areas.

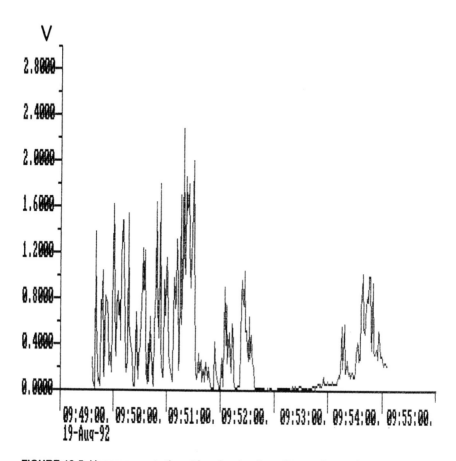

FIGURE 10-5. Vapor concentration at face level a given distance from solvent (methyl ethyl ketone) being applied on a sheet of aluminum in a classroom. 1 V = approximately 45 ppm. Distance from the solvent to the sensor: (1) 0.5 m; (2) 5.0 m; (3) 1.5 m; (4) 1.0 m.

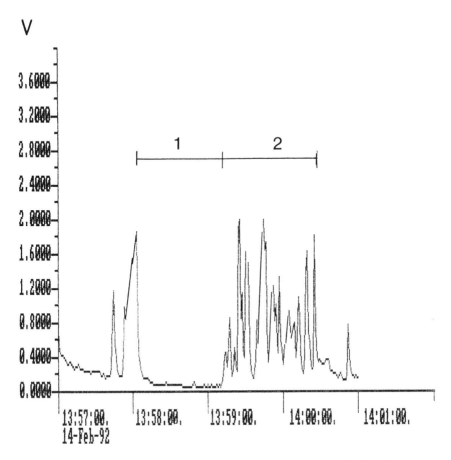

FIGURE 10-6. Ambient concentration of N-butyl acetate vapor under the hood in a classroom. 1 V is approximately 35 ppm N-butyl acetate. (1) Main door open; (2) main door closed.

CHAPTER 11

Further Possibilities of Sensor–Logger Combinations

From the examples presented in the preceding chapters, it is evident that it is possible, with the range of sensors and loggers now available, to conduct quite elaborate on-the-spot surveys of a wide variety of environmental stresses in modern industry, and, to some extent, the effect which some of these stresses have on the working individual. But this is only the beginning. New types of sensors as well as loggers are steadily appearing on the market, and additional ones can easily be made to meet specific requirements.

Most of the examples introduced in this presentation are based on the Squirrel logger (Grant Instruments, Ltd., Cambridge, England), simply because most of my own experience is based on this particular logger. There are, however, a number of other loggers available, and others no doubt will appear in the future. One of the loggers which has recently appeared on the market is the 12-bit Ramlog EI 9000 (produced by a.b.i. data, Brussels, Belgium). It has up to 16 analog channels, high memory capacity, and may handle from 100 samples per second to one sample per hour. It may be used as a meter as well as a logger, or both. It uses disposable or rechargeable batteries. It has output for a printer or for a computer. It weighs 500 g and its size is 18 × 10 × 4.5 cm.

A complete system for the measurement of an electromyogram (EMG) and postural angle is the so-called Physiometer PHY-400 (produced by Premed Ltd., Oslo, Norway). In addition to EMG and postural angles, other parameters, such as temperature can be included.

In addition to the range of sensors described in the previous chapters, there are a great variety of additional applicable sensors available for applied work physiology in the field. A personal heat stress monitor, the Metrosonics hs-383 (produced by Metrosonics Inc., General Products Division, Rochester, NY, U.S.A.) is for the purpose of ambulatory logging of temperature and heart rate. Its size is 7.6 × 2.0 × 12.6 cm, and it weighs 300 g.

A wide selection of temperature probes, both thermistors and thermocouples with different cable length, is supplied by Grant Instruments, Ltd., Cambridge,

England, both for the logging of environmental heat stress, and the resulting heat strain on the exposed individual. This company also supplies humidity probes, such as the VH-L probes made by Vaisala in Finland, and VH-H probes which are made by Lee-Integer in England, both of which are compatible with the Grant recording systems.

A wide spectrum of gas monitors, compatible with the described ambulatory loggers, is supplied by Metrosonics, Inc., Rochester, NY, U.S.A.

The Swiss-made MiniAir 6 vane anemometer (Schiltknecht Messtechnik AG) is a handy sensor for flow-velocity and temperature measurement, which in combination with a logger can be used for an assessment of indoor environments, or for ambulatory assessment of air velocity, draft, etc.

Over the years, a number of systems have been developed for the purpose of measuring the oxygen uptake, and hence the energy expenditure in subjects exercising in the field. Eley et al., (1978) have described a respirometer system for use in the field for the measurement of oxygen uptake—known as the Miser—which consists of a photoelectronic gas meter, a volume counter, and a vacuum sampler. It weighs only about 600 g and is powered by rechargeable cells which will last for some 8 hours without recharging. The disadvantage with this system is that it does not analyze the expired air, but takes and stores a sample of it for subsequent analysis in the laboratory.

With the sophisticated electronic sensors and loggers now available, it is possible to construct tailor-made systems which continuously log the expiratory flow volume, its oxygen and carbon dioxide content, temperature, and even heart rate at the same time, from the same subject, in the field.

The sensitivity of the blood pressure to mental and physical stress and the value of ambulatory blood pressure measurement as a cardiovascular reaction to stress have been emphasized earlier in this book. The development of new techniques involving noninvasive ambulatory monitoring of blood pressure has made it possible to obtain readings of blood pressure at regular intervals in patients and healthy individuals while they go about their normal daily activities. This has made it possible to relate blood pressure variations to physical and mental activity, as well as to emotional and psychosocial factors.

Pickering (1990), in his review of some of the physiological aspects of noninvasive ambulatory blood pressure monitoring, describes the basic principles behind some of the fully automatic devices which have been developed, and which can be programmed to allow for the frequency of readings to be preset. An electric pump or a CO_2 cartridge inflates the cuff at fixed intervals. They may utilize the Korotkoff sound technique with a piezoelectric microphone placed under the cuff on the upper arm, or the oscillometric technique, measuring the oscillation of pressure within the cuff. According to Pickering (1990) both techniques work reasonably well, but they are susceptible to artifacts if the subject is moving the arm when the blood pressure is being measured. Since the subject is able to feel the increasing pressure of the cuff as it is being inflated, he or she should be told to keep the arm still while the reading is being taken. As pointed out by Pickering, technical improvements are in steady progress. The latest version weighs less than a kilogram and allows for more than 100 blood pressure readings to be taken in 24 hours. One such product is the PAR physio-port ambulatory blood pressure recorder (produced by PAR-Electronic GmbH in Berlin, Germany). It weighs 980 g with battery. Another is the Takeda medical ambulatory blood pressure monitoring system TM-2420/TM-2000, provided by AND Corporation, Inc., Milpitas, CA, U.S.A. It weighs even less than the PAR physio-port recorder.

Sunderberg (1987) monitored the blood pressure for 24-hour periods in 9 healthy normotensive subjects with an ambulatory, automatic, noninvasive blood pressure recorder, the Pressurometer III, produced by Del Mar Avionics. This instrument allows for automatic blood pressure readings at 7.5, 15, and 30 min intervals for an entire 24-hour period. The recorder is attached to the subject's belt and is connected to a conventional blood pressure cuff wrapped around the upper arm. The microphone sensing the pulse is taped in position over the brachial artery and under the cuff, which is automatically inflated by a pump in the recorder at preset intervals. The microphone than auscultates the Korotkoff sounds as the cuff is deflated. The results were compared with simultaneous hourly measurements of the blood pressure by the conventional auscultatory method. There were no statistically significant differences between the automatic and the auscultatory readings. Here again the automatic method's sensitivity to motion artifacts was found to be a disadvantage. This drawback, however, might be overcome by devoting more attention to the instruction and training of the subject, and perhaps by incorporating a filtering mechanism for such artifacts in the instrument.

Another parameter which is often an occupational health problem and which can easily be recorded by existing sensors is vibration both whole body vibration and hand/arm vibration. Brüel and Kjær in Copenhagen, Denmark have a product known as the Brüel & Kjær Human Vibration Meter 2512. The real problem with vibration, however, is not so much to sense it, but to do something about it, preferably to eliminate it, but in any case to prevent its harmful effects.

In the case of noise, there are excellent noise monitoring instruments available for the visualization of its kind and magnitude, such as the Brüel & Kjær Noise Dose Meter Type 4436, or the Du Pont Mark Series Audio dosimeter system. The problem is that it takes a certain amount of discipline and determination to generate an atmosphere in an industrial work place which makes everyone use the necessary protective measures as a matter of course.

The powerful magnetic fields encountered in certain types of electrolytic industries, such as the aluminum, zinc, and other metallurgic plants, have been a matter of concern for a number of years. There are suitable sensors available for the assessment of both static and variable magnetic fields, such as the AMS-prototype 3D combined magnetic sensor developed by the Center for Industrial Research in Oslo, Norway. Such logger-compatible sensors make it possible to map the level and location of the fields in question. But here again, the problem is to detect any convincing evidence of objective pathological effects of such magnetic fields on the people exposed to them.

From this brief review, it is evident that there is a wide spectrum of sophisticated electronic sensors and loggers available for the survey and control of the common environmental stresses, and for the protection of those who are exposed to them. At the same time the trend in our industrialized countries is for the responsible authorities to introduce rules and regulations aimed at making industry itself responsible for the survey and control of their own environments, and for the protection of their workers. It is with this trend in mind that one would expect to see a greater eagerness on the part of industry to start using some of these sensor–logger systems internally in order to enhance their insight into their own environmental problems, as a basis for action leading to improvements.

This can be done by a combination of the use of some of these sensor–logger systems for ambulatory assessment of individual exposure to environmental stresses during work, and for the transfer via remote control (Modem) of the readings from local sources in the plant to computers in central offices for display of the readings for the benefit of those who are involved in the health and safety and for the productivity of the plant.

A continuous visualization of pertinent parameters which are subject to common concern for both workers and management is now possible. This fact may tend to bring all factions of the plant together in front of the computer screen for an objective discussion of the displayed facts, in an endeavor to improve the working environment, the health and well being of the worker, and thereby also improve productivity for the good of all.

References

Accorsi, A. and P. Hure: Detecteurs portatifs d'hydrogene sulfure. Cathiers de notes documentaires *138(1)*:31-39, 1990.

Adolph, E.F. and associates: Physiology of man in the desert. Interscience Publishers, New York, 1947.

Alm, N.O. and K. Rodahl: Work stress in Norwegian coal miners in Spitsbergen (unpublished) 1979.

Åstrand, I., P. Fugelli, C.G. Karlsson, K. Rodahl, and Z. Vokac: Energy output and work stress in coastal fishing. Scand. J. Clin. Lab. Invest. *31*:105, 1973.

Åstrand, P.-O. and K. Rodahl: "Textbook of Work Physiology", 1st ed. McGraw-Hill, New York, 1970.

Åstrand, P.-O. and K. Rodahl: "Textbook of Work Physiology", 3rd ed. McGraw-Hill, New York, 1986.

Baker, M.A.: Brain cooling in endotherms in heat and exercise. Am. Rev. Physiol., *44*:85, 1982.

Basmajian, J.V.: "Muscles Alive", Williams & Wilkins, Baltimore, 1979.

Bedi, J.F. and S.M. Horvath: Inhalation route effects on exposure to 2.0 parts per million sulfur dioxide in normal subjects. J.A.P.C.A., *39(11)*:1448-1452, 1989.

Benzinger, T.H. and G.W. Taylor: Cranial measurements of internal temperature in man, in J.D. Hardy (Ed) "Temperature: Its Measurement and Control in Science and Industry", Vol. 3, Part 3, p. 111. Reinhold, New York, 1963.

Beshir, M.Y., J.D. Ramsey, and C.L. Burford: Threshold values for the Botsball: A field study of occupational heat. Ergonomics, *25(3)*:247, 1982.

Bigland, B. and O.C.J. Lippold: The relation between force, velocity and integrated electrical activity in human muscles. J. Physiol., *123*:214, 1954.

Bjørklund, R.A., A.-H. Kulsrud, and K. Rodahl: The effects of protective helmets for pedal cyclists in terms of heat stress of the head, and psycho-motor performance. Nat. Inst. Occup. Health, Oslo, Norway, unpublished, 1991.

Blix, I., A. Bolling, and K. Rodahl: En arbeidsfysiologisk undersøkelse ombord på M/S Vindafjord på fart fra Afrika til Europa, 3S-rapportserien, Oslo, 1979.

Botsford, J.H.: A wet globe thermometer for environmental heat measurement. AIHA J., *32*:1, 1971.

Bottolfsen, T.: The effect of acupuncture on muscle tension, as evidenced by EMG registration. Thesis for acupuncture qualifying examination, Kristiansand, 1991.

Brouha, L.: "Physiology in Industry". Pergamon Press, New York, 1960.

Brown, G.A. and G.M. Williams: The effect of head cooling on deep body temperature and thermal comfort in man. Aviation, Space, and Environmental Medicine, 583, 1982.

Burton, A.C. and O.G. Edholm: "Man in a Cold Environment", Edward Arnold Publishers, London, 1955.

Cabanac, M.: Keeping a cool head. News in Physiological Sciences, *1*:41, 1986.

Chung, K.Y.K. and N.P. Vaughan: Comparative laboratory trials of two portable direct-reading dust monitors. Ann. Occup. Hyg., *33(4)*:591-606, 1989.

Ciriello, V.M. and S.H. Snook: The prediction of WBGT from the Botsball. Am. Ind. Hyg. Assoc. J. *38*:264, 1977.

Cunxin, Y., J. Xing, and Y. Changying: observations on clinical therapeutic effect in treating soft tissue injuries by acupuncture, with pain threshold and electromyography as parameters. J. Trad. Chin. Med., *9(1)*:40-44, 1989.

Dernedde, E.: A correlation of the Wet-Bulb Globe Temperature and Botsball Heat Stress Indexes for Industry. Amer. Ind. Hyg. Assoc. J. *53(3)*:169-174, 1992.

Dykes, R.W.: Factors related to the dive reflex in harbour seals: Respiration, immersion bradychardia and lability of the heart rate. Can. J. Physiol. Pharmacol. *52*:259-262, 1974.

Egeland, T., P. Lossius, N. Enger, N. Gundersen, A. Bolling, and E. Jebens: Varmebelastningsundersøkelse ved Fiskaa Verk, Kristiansand. Arbeidsfysiologisk Institutt, Oslo, AFYI Publikasjon, 1981.

Eley, C., R. Goldsmith, D. Layman, G.L.E. Tan, E. Walker, and B.M. Wright: A respirometer for use in the field for the measurement of oxygen consumption. "The Miser", a miniature, indicating and sampling electronic respirometer. Ergonomics *21(4)*:253-264, 1978.

Erikssen, J., K. Knudsen, P. Mowinckel, T. Guthe, J.P. Lützow Holm, R. Brandtzæg, and K. Rodahl: Blodtrykksstigning hos stresseksponerte industriarbeidere. Tidsskr. Nor. Lægeforen. nr. 22, *110*:2873-7, 1990.

Floyd, W.F. and A.T. Welford: Symposium on Fatigue. H.K. Lewis & Co., London, 1953.

Floyd, W.F. and P.H.S. Silver: The function of the erectores spinae muscles in certain movements and postures of man. J. Physiol. *129*:184, 1955.

Greenleaf, J.E. W. van Beaumont, P.J. Brock, L.D. Montgomery, J.T. Morse, E. Shvartz, and S. Kravik: Fluid–electrolyte shifts and thermoregulation: rest and work in heat with head cooling. Aviation, Space and Environmental Medicine, *51(8)*:747-53, 1980.

Guthe, T., K. Gundersen, H. Skjennum, L.D. Klüwer and K. Rodahl: En orienterende undersøkelse av varmestress og muskelbelastning ved utvalgte arbeidsplasser ved Hadeland Glassverk A/S. KIL-amil-dok-4, 25 Sept. 1990.

Guthe, T., J. Meyer, and K. Rodahl: En orienterende logging av svoveldioksyd innholdet i arbeidsatmosfæren ved Arendal Svelte verk, 11-13 mars 1991. Upublisert rapport, 1991.

Hagberg, M.: On evaluation of local muscular load and fatigue by electromyography. Arbete och Hälsa. Vetenskapelig Skriftserie, Arbetarskyddsverket, Stockholm, Sweden, 1981.

Heldal, K., E. Melbostad, B. Tvedt, W. Eduard, A. Skogstad, P. Søstrand, E. Bye, P. Sandven, and J. Lassen: Helse og arbeidsforhold ved behandling av kommunalt avløpsvann. H.D. 1024/91 FOU, Nat. Inst. Occup. Health, Oslo, Norway, 1992.

Hettinger, Th. and K. Rodahl: A work physiology study of an assembly line operation. J. Occup. Med. *2(11)*:532-535, 1960.

Hurwitz, B.E. and J.J. Furedy: The human dive reflex: an experimental, topographical and physiological analysis. Physiol. Behav. *36*:287-294, 1986.

Irving, J.M., E.C. Clark, I.K. Crombie, and W.C.S. Smith: Evaluation of a portable measure of expired-air carbon monoxide. Prev. Med. *17*:109, 1988.

Jahr, J., T. Norseth, and K. Rodahl: Sammenligning av arbeidsforholdene i to typer elektrolysehaller på Sunndalsøra. Arbeidsforskningsinstituttene, Oslo, AFYI publikasjon, 1971.

Jansen, T., B. Olsen, M. Hatlevold, N. Enger, T. Guthe, H. Johannessen, and K. Rodahl: En undersøkelse av varme-eksponeringen ved Lista Aluminiumverk. Arbeidsfysiologisk Institutt, Oslo, AFYI publikasjon, 1982.

Jarvis, M.J., M.A.H. Russell, Y. Saloojee: Expired air carbon monoxide: a simple breath test of tobacco smoke intake. Brit. Med. J. *281*:484, 1980.

REFERENCES

Jarvis, M.J., B. Belcher, C. Vesey, and D.C.S. Hutchison: Low cost carbon monoxide monitor in smoking assessment. Thorax, *41*:886, 1986.

Kangas, J., A. Nevalainen, A. Manninen, and H. Savolainen: Ammonia hydrogen sulfide and methyl mercaptides in Finnish municipal sewage plants and pumping stations. The Science of the Total Environment, *57*:49-55, 1986.

Kawakami, Y., B.H. Natelson, and A.B. DuBois: Cardiovascular effects of face immersion and factors affecting diving reflex in Man. J. Appl. Physiol. *23*:964-970, 1967.

Keatinge, W.R. and R.E.G. Sloan: Deep body temperature from aural canal with servo-controlled heating to outer ear. J. Appl. Physiol. *38*:919-921, 1975.

Kloetzel, K., A. Etelvino do Andrade, J. Falleiros, and J. Cota Pacheco: Relationship between hypertension and prolonged exposure to heat. J. Occup. Med. *15(11)*:878-80, 1973.

Konz, S. and V.K. Gupta: Water cooled hood affects creative productivity. Ashrae J., July, *11(7)*:40-43, 1969.

Lammert, O.: Maximal aerobic power and energy expenditure of Eskimo hunters in Greenland. J. Appl. Physiol. *33*:284, 1972.

Lehmann, G.: Praktische Arbeitsphysiologie. Georg Thieme Verlag, Stuttgart, 1953.

Lilienfeld, P. and R. Stern: Personal dust monitor–light scattering. U.S. Bureau of Mines Contract Report No. H0308132, U.S. Dept. of the Interior, Pittsburgh, U.S.A., 1982.

Maehlum, S., A. Bolling, P.O. Huser, E. Jebens, and O. Tenfjord: En arbeidsfysiologisk undersøkelse av varmebelastningen ombord i M/S Tarn i fart på den Persiske Gulf. 3S-Rapportserien, Oslo, 1978.

Magnus, P., A. Bolling, I. Eide, N. Enger, N. Gundersen, T. Guthe, E. Jebens, K.E. Knudsen, and K. Rodahl: Varmestress i ferrolegeringsindustrien. En orienterende undersøkelse ved Fiskaa Verk, Kristiansand. Arbeidsfysiologisk Institutt, Oslo, AFYI Publikasjon, 1980.

Maltun, K.R.: EMG målinger under utskrivning av forsikringspoliser, personal communication, 1990.

Martin, H.deV. and S. Callaway: An evaluation of the heat stress of protective face mask. Ergonomics *17*:221, 1974.

Morales, I.V. and S. Konz: The physiological effect of a water cooled hood in a heat stress environment. ASHRAE Trans. *74(2)*:49, 1968.

Mundal, R., J. Erikssen, R.A. Bjørklund, and K. Rodahl: Elevated blood pressure in air traffic controllers during a period of occupational conflict. Stress Medicine *6:*141-144, 1990.

Natusch, D.F.S. and B.J. Slatt: Hydrogen sulfide as an air pollutant, in Air Pollution Control, Part III. Measuring and Monitoring. Straus, New York, 1978, Chap. 9.

Nes, H., R. Karstensen, and K. Rodahl: Varmestressundersøkelse ved Elkem Aluminium, Mosjøen, Technical report, 1990.

Nes, H., R. Karstensen, and K. Rodahl: Måling av hjertefrekvens i romtemperatur og under varmebelastning i Søderberghall. Elkem Aluminium Mosjoen Laboratorierapport, januar 1991a.

Nes, H., R. Karstensen, and K. Rodahl: Måling av den fysiske arbeidsbelastning hos operatører som er i varme jobber ved Elkem Aluminium, Mosjøen. Elkem Aluminium Mosjøen laboratory report, May 1991b.

Newburgh, L.H. (Ed.): "Physiology of Heat Regulation and the Science of Clothing". W.B. Saunders, Philadelphia, 1949.

Nilsson, S., P. Lereim, M. Braaten, I. Greger, P.O. Huser, and K. Rodahl: Undersøkelse av arbeidsbelastningen ved Flensdrivaksellinjen, Kongsberg Våpenfabrikk, januar/februar 1970. Arbeidsforskningsinstituttene og Kongsberg Våpenfabrikk. unpublished, 1970.

Nitter, L.H., J. Grønli, A. Lie, L.D. Klüwer, T. Guthe, and K. Rodahl: Varmestress ved PLM Moss Glassverk. En orienterende undersøkelse av ulike arbeidsplasser. februar–mars, 1990. KIL-amil-dok-5. March 29th, 1990.

Norman, L.A.: Mouse joint — another manifestation of an occupational epidemic? West. J. Med. *155*:413-415, 1991.

Nunneley, S.A., P. Webb, and S.J. Troutman: Head cooling during work and heat stress. Ergonomics *13(4)*:527, 1970.

Pepler, R.D.: Performance and Well-Being in Heat, In J.D. Hardy (ed): "Temperature; Its Measurement and Control in Science and Industry". Vol. 3, part 3, p. 319. Reinhold Book, New York, 1963.

Peterson, J.E.: Postexposure relationship of carbon monoxide in blood and expired air. Arch. Environ. Health. *21*:172, 1970.

Philipson, L., R. Sörbye, P. Larsson and S. Kaladjev: Muscular load levels in performing musicians as monitored by quantitative electromyography. "Medical Problems of Performing Artists", Hanley & Belfus, Philadelphia, 1990.

Pickering, T.G.: Physiological aspects of noninvasive ambulatory blood pressure monitoring. News in Physiological Sciences (NIPS), *5*:176-179, 1990.

Ramanathan, L.N.: A new weighing system for mean surface temperature of the human body. J. Appl. Physiol. *19(3)*:531, 1964.

Rea, J.N., P.J. Tyrer, H.S. Kasap, and S.A.A. Beresford: Expired air carbon monoxide, smoking, and other variables. A Community study. Brit. J. Prev. Soc. Med. *27*:114, 1973.

Rees, P.J., C. Chilvers, and T.J.H. Clark: Evaluation of methods used to estimate inhaled dose of carbon monoxide. Thorax, *35*:47, 1980.

Riggs, C.E., D.J. Johnson, B.J. Konopka, and R.D. Kilgour: Exercise heart rate response to facial cooling. Eur. J. Appl. Physiol., *47*:323, 1981.

Rodahl, A.: Muskelspenning og feilbelastningslidelser. Boen Parkettfabrikk A/S, Kristiansand, 1989.

Rodahl, K.: The toxic effect of polar bear liver. Norsk Polarinstitutt Skrifter No. 92, 1949.

Rodahl, K.: U.S. Air Force Survival Ration Studies in Alaska. Arctic, *3*:124, 1950.

Rodahl, K.: Eskimo metabolism. A study of racial factors in basal metabolism. Norsk Polarinstitutt Skrifter, no. 99, 1954.

Rodahl, K.: Regelmessig tilførsel av væske har betydning for utholdenhet. Arbeidsgiveren, *No. 24*:556, 1975a.

Rodahl, K.: Arbeidsfysiologiske undersøkelser ved Jøtul, Oslo. Arbeidsfysiologisk Institutt, Oslo, AFYI Publikasjon, 1975b.

Rodahl, K.: En arbeidsfysiologisk undersøkelse av lasting og lossing av fly på Fornebu. I-II. Arbeidsfysiologisk Institutt, Oslo, AFYI Publikasjon, 1976.

Rodahl, K.: Arbeidsstress til sjøs. System for Sikkert Skip(3S) NTNF Rapport 80/2. Norges Teknisk-Naturvitenskapelige Forskningsråd, Oslo, 1980.

Rodahl, K.: Heat Stress: Norwegian experience, in J.P. Hughes (ed.): Health Protection in Primary Aluminium Production. Vol. 2. International Primary Aluminium Institute, London, 1981.

Rodahl, K.: "The Physiology of Work", Taylor & Francis, London, 1989.

Rodahl, K. and T. Moore: The vitamin A content and toxicity of bear and seal liver. Biochem. J. *37*:166, 1943.

Rodahl, K., S.M. Horvath, N.C. Birkhead, and B. Issekutz, Jr.: Effects of dietary protein on physical work capacity during severe cold stress. J. Appl. Physiol. *17*:763, 1962.

REFERENCES

Rodahl, K., N.C. Birkhead, B. Issekutz, Jr., J.J. Blizzard, G.J. Haupt, R.N. Meyers, P.A. Lachance, and E.W. Speckmann: Effects of prolonged bed rest. Conference on nutrition in space and related waste problems. National Aeronautics and Space Administration, NASA, SP-70, 1964.

Rodahl, K., Z. Vokac, and K. Johnsen: Rapport angående undersøkelse foretatt ved Sande Paper Mill A/S. Arbeidsfysiologisk Institutt, Oslo, AFYI Publikasjon, 1971.

Rodahl, K., Z. Vokac, P. Fugelli, O. Vaage, and S. Maehlum: Circulatory strain, estimated energy output and catecholamine excretion in Norwegian coastal fishermen. Ergonomics 17(5):585, 1974.

Rodahl, K. and P.O. Huser: En arbeidsfysiologisk vurdering av sugevognoperasjon i elektrolyseavdelingen på magnesium-fabrikken ved Norsk Hydro, Hærøya. Arbeidsfysiologisk Institutt, Oslo, AFYI Publikasjon, sept. 1976.

Rodahl, K. and Z. Vokac: Work stress of Norwegian trawler fishermen. Ergonomics 20(6):633, 1977a.

Rodahl, K. and Z. Vokac: The physiology of fishing. Nordic Council Arct. Med. Res. Rep. 18:22, 1977b.

Rodahl, K., H. Ramstad, R. Mundal, P.O. Huser, and B. Jarmark-Robertsson: Flygelederne ved Oslo Lufthavn, Fornebu. En arbeidsfysiologisk undersøkelse. Arbeidsfysiologisk Institutt, Oslo, AFYI Publikasjon, 1981.

Rodahl, K., P.O. Huser, and B. Jarmark-Robertsson: Varmestress og arbeidsbelastning hos tappere ved Odda Smelteverk. Arbeidsfysiologisk Institutt, Oslo, AFYI Publikasjon, 1984.

Rodahl, K., R. Mundal, P.O. Huser, and P. Bjørgan: Leder for små og mellomstor bedrifter; en arbeidsfysiologisk undersøkelse. Arbeidsfysiologisk Institutt, Oslo, AFYI Publikasjon, 1985.

Rodahl, K. and T. Guthe: Physiological limitations of human performance in hot environments. with particular reference to work in heat-exposed industry. Chapter 2 in: Mekjavic et al. (Ed.): Environmental Ergonomics. Taylor & Francis, London, 1988.

Rodahl, K., T. Guthe, L.H. Nitter, and L.D. Klüwer: Ambulatorisk minicomputer–logging av varmestress ved Moss Glassverk. En orienterende undersøkelse. Technical Report. GKV-amil-dok-2, sept. 1989.

Rodahl, K. and T. Guthe: Praktisk mätning och praktisk bruk av EMG. Arbete mænniska miljö, 2.90:116, Stockholm, 1990.

Rodahl, K. and T. Guthe: Beskyttelse av hodet mot varmestress. KIL-amil-dok-12, 18 november 1991.

Rodahl, K., T. Guthe, L.D. Klüwer, and J. Meyer: Beskyttende effekt av reflekterende bekledning mot strålevarme ved Moss Glassverk. KIL-amil-dok-6., 15 november 1990.

Rodahl, K., T. Guthe, and L.D. Klüwer: Bruk av viftehjelm på varme arbeidsplasser. En orienterende undersøkelse. KIL-amil-dok-14. 23 desember 1991a.

Rodahl, K., T. Guthe, L.D. Klüwer, and J. Meyer; Orienterende undersøkelse med sensorlogging og kontinuerlig registrering av løsemiddeldamp-konsentrasjoner ved Scandia Kjemiske A/S, Oslo, KIL-amil-dok-10, 25 oktober, 1991b.

Rodahl, K., T. Guthe, and L.D. Klüwer: Virkningen av regelmessig og tilstrekkelig væsketilførsel hos operatører ved PLM Moss Glassverk A/S. KIL-amil-dok-13, 17 desember 1991c.

Rodahl, K., T. Guthe, and L.D. Klüwer: En orienterende undersøkelse av muskelspenningen hos operatører på utvalgte arbeidsplasser ved Dynoplast A/S, Kongsvinger. KIL-amil-dok-9, 12 oktober, 1991d.

Rodahl, K., L. Leuba, L. Klüwer, and T. Guthe: Varmestress ved Moss Glassverk. Hodets rolle i den umiddelbare økning av hjertefrekvensen ved plutselig eksponering for varme. PIL-amil-dok-18, 8 oktober 1992a.

Rodahl, K., T. Guthe, and L.D. Klüwer: En orienterende undersøkelse av mekanismen for økningen i hjertefrekvensen ved eksponering til varme. KIL-amil-dok-17, 19 juni 1992b.

Rodahl, K., T. Guthe, L.D. Klüwer, L. Leuba: Beskyttelse mot varmestress ved PLM Moss Glassverk. En innledende, orienterende undersøkelse. KIL-amil-dok-16, 10 april 1992c.

Rodahl, K., T. Guthe, L.D. Klüwer, and T. Eklund: Logging av muskelspenning i arm-og skuldermuskler ved bruk av MUS og PC. Project rapport, Norsk Hydro, Oslo, 1992d.

Rodahl, K., R.A. Bjørklund, A.-H. Kulsrad, L.D. Klüwer, and T. Guthe: Effects of protective helmets on body temperature and psycho-motor performance. The Fifth International Conference on Environmental Ergonomics, Maastricht, Nov. 2-6, 1992e.

Shvartz, E.: Effect of a cooling hood on physiological responses to work in a hot environment: J. Appl. Physiol., $29(11)$:36, 1970.

Shvartz, E.: Effect of neck versus chest cooling on responses to work in heat. J. Appl. Physiol. $40(5)$:668, 1976.

Søstrand, P.: Sewage gases in Norwegian plants and pumping stations. Nat. Inst. Occup. Health, Oslo, Norway. unpublished, 1992.

Sundberg, S.: Noninvasive, automatic 24-h ambulatory blood pressure monitoring in normotensive subjects. Eur. J. Appl. Physiol. 56:381-383, 1987.

Thompson, R.S., F.P. Rivara, and D.C. Thompson: A case-control study of the effectiveness of bicycle safety helmets. New Engl. J. Med. $320(21)$:1361-1367, 1989.

Tola, S., H. Rütiimäki, T. Videman, E. Viikari, and K. Hänninen: Neck and shoulder symptoms among men in machine operating, dynamic physical work and sedentary work. Scand. J. Work Environ. Health, 14:299, 1988.

Veiersted, K.B.: The reproducibility of test contractions for calibration of electromyographic measurements. Eur. Appl. Physiol. 62:91-98, 1991.

Vokac, Z., and K. Rodahl: A field study of rotating and continuous night shifts in a steel mill. In "Experimental Studies of Shiftwork", Forskungsberichte des Landes Nordrheim–Westfalen, Nr. 2513, 168-173, Westdeutscher Verlag, 1975.

Vokac, Z. and K. Rodahl: Maximal aerobic power and circulatory strain in Eskimo hunters in Greenland. Nordic Council Arct. Med. Res. Rep., No. 16:16, 1976.

Wald, N.J., M. Idle, J. Boreham, and A. Bailey: Carbon monoxide in breath in relation to smoking and carboxyhaemoglobin levels. Thorax, 36:366, 1981.

Westgaard, R.H., M. Wærsted, T. Jansen, and K. Korsund: Belastninger og belastningslidelser hos produksjonsarbeidere ved Helly Hansen A/S. Arbeidsfysiologisk Institutt, Oslo, AFYI Publikasjon, 1984.

Westgaard, R.H., P.O. Huser, B. Jarmark-Robertsson, and K. Rodahl: Fysisk belastning for catering-personalet på Statfjord-feltet i: Westgaard et al.: "Arbeidsmiljø, belastningslidelser og sykefravær blandt forpleieningspersonell på Statfjordfeltet". Universitetsforlaget, Oslo, 1987.

Wyon, D.P., I. Andersen, and G.R. Lundqvist: The effects of moderate heat stress on mental performance. Scand. J. Work Environ. Health, 5:352, 1979.

Zijlstra, W.G., A. Buursma, and A. Zwart: Performance of an automated six wavelength photometer (Radiometer OSM3) for routine measurement of hemoglobin derivations. Clin. Chem. $34(1)$:149, 1988.

Index

Acclimatization, to heat, 12
Air pollution, 9
Aircraft pilots, stress of, 31–32
Airport luggage handling, stress of, 37
Air traffic controllers:
    blood pressure, 36
    resting heart rate, 36
    stress, 33
Aluminum production, 9, 13
Ambulatory loggers, 39–40
Arctic trappers, 1
Assembly line operation, 3
    productivity, 4
    stress, 22

Bed rest, 3
Blood pressure, effect of occupational conflicts, 36
Botsball thermometer, 9, 92

Calcium carbide production plant, 15
Carbon monoxide:
    in a silicon carbide production plant, 127
    in tobacco smokers, 133
    logging of, 125–134
    uptake of, 125–126, 130–134
Catch-handlers, energy expenditure, 24
Catering offshore, work load, 31
Cement factory, 14
Chemical vapors, in the working atmosphere, 145–151
Circadian rhythm, 20
Cockpit environment, 32
Cold stress:
    in arctic coal miners, 16
    in industry, 16
    rectal temperature, 17
    skin temperature, 17

Douglas bag method, 7
Dust, sensing of, 121–123

Eltek Special, heart rate recorder, 60–62
Environmental temperature:
    effect of, 91
    indices, 93–94
Eskimos, 2, 26
    seal hunting, 26–27

Fishing:
    circulatory strain, 25
    energy cost, 23–24
    stress, 22–24
    urinary stress hormones, 24
Fluoride exposure, 9
Forced expiratory volume (FEV), 9
    tobacco smoking, 9

Heart rate:
    aluminum plant operations, 54–55
    and work load, 5, 49
    effect of heat stress, 55–60
    ferroalloy plant operations, 51–52
    foreign exchange agent, 54
    recording of, 49–50, 60–62
    simulated bank robbery, 52–54
Heat strain, 9
    assessment of, 94–97
    in aluminum production, 97–99
    in glass production, 99–105
Heat stress, 9–10, 16
    accidents, 100
    effect of fluid intake, 14, 105–106
    fluid balance, 102
    heart rate, 55–60
    hemodilution, 13
    hypohydration, 98
    in the aluminum industry, 97–99
    in the glass industry, 99–105
    monitor of, 91–120
    protection, 103
    protection of the head, 111–114
    protective clothing, 106–111
    rectal temperature, 13
    skin temperature, 13
    sodium concentrations in the blood, 13
    sweat rates, 13
Humidifier, effect of, 121
Humidity, relative, 121
    sensing of, 121–124
Hydration, state of, 7
Hydrogen sulfide:
    effects of pumping sludge, 142–143
    logging of, 141–144
Hypertension, and industrial heat exposure, 10–13

Industrial physiology:
  in the U.S., 3
  in Norway, 4

Loggers, 40, 153
Longline fishing, 24

Muscle tension:
  biofeedback, 89–90
  calibration, 66
  effect of acupuncture, 79–84
  effect of Mouse-based PC operations, 64–85
  effect of ergonomic improvements, 63
  extensor digitorum muscle, 87
  in floorboard production, 74–77
  integrated EMG, 66
  logging of, 63–89
  Myolog recording, 66, 68–77
  neuromuscular complaints, 64, 77–89
  plastic container production, 69–74
  symptoms of, 63
  trapezius muscle, 69

Organic vapors:
  in aircraft maintenance, 149
  in a paint plant, 145
  in the working atmosphere, 145–151

Paper mill, 15
Physical fatigue, 6, 8
Physical work:
  capacity, 3
  factors affecting, 42
  load, 8
  (see also Work load)
Polar bear liver, vitamin A content and toxic effect of, 1
Protective helmets:
  effect of, 115–120
  psychomotor performance, 119
  temperature of the head, 119
Pulmonary function, 9
Purse seine fishing, 25
Radioelectrocardiography, 7
Rectal temperature, recording of, 95–96
Renin, plasma concentrations, 12

Sailors:
  blood pressure, 28
  heat stress, 28, 30

Sailors (continued):
  heart rate, 28
  stress, 27
  physical fitness, 28
  work load, 28
Sensors:
  anemometer, 154
  blood pressure, 154–155
  carbon monoxide, 125–134
  carbon monoxyhemoglobin, 126–130
  chemical vapors, 145
  dust, 121–124
  gas monitors, 154
  heart rate counting, 49
  heat, 91–97
  heat stress monitor, 153
  humidity, 121–124
  hydrogen sulfide, 141–144
  magnetic fields, 155
  muscle tension, 66
  noise, 155
  organic vapors, 145
  oxygen uptake, 154
  sulfur dioxide, 135–140
  vibration, 155
Shift work:
  circadian rhythms, 19
  physiological effects, 18
Skin temperature, recording of, 96–97
Small business managers, stress of, 38
Speed skating, 5
Squirrel meter/logger, 40–42
  sensors for, 44–46
Sulfur dioxide:
  effects of, 135
  in silicon carbide production, 135–140
  logging of, 135–140
Survival rations, 1

Trawler fishing, 24

Wet Bulb–Globe Temperature Index (WBGT), 9, 91, 93–94
Wet Globe Temperature Index (WGT), 9, 91, 93–94
Work load:
  capacity, 6
  heart rate, 5, 49
  oxygen uptake, 7
  relative, 7
  (see also Physical work)